The Ancients Come Forth

The Ancients Come Forth
Lessons from the Guardians

Richard Spencer

Pensacola, Florida

Copyright © 1999 by Richard Spencer

All rights reserved. No part of this book may be reproduced by any means or in any form whatsoever without written permission from the publisher, except for brief quotations embodied in literary articles or reviews.

Contents printed on recycled, acid-free paper.

Printed in the United States of America
First Printing: March, 1999
10 9 8 7 6 5 4 3 2 1

Library of Congress Catalog Number: 98-68101

ISBN: 0-9656495-1-2

Openings, P.O. Box 2430-1356, Pensacola, FL 32513

To Alma, my love

The Ancients Come Forth

1

The pattern of cosmic movement that informs itself moved out to the edges of the galaxy. Seeking to know itself more, the pattern touched the planet, now called Earth. It reached out at the same moment and encircled a binary star called the Dog Star - two stars circling around each other, each contributing opposite forces to the dance, forces in tension that fed growth and diversity.

In an instant, material formed from the swirl of opposites and was thrust out along the connection that had been created by the pattern of movement. The comet streaked across dark space and plummeted into the ocean of Earth. Where there were no opposing forces before, now there were. This was the Great Seeding.

The planet herself, shocked by the intrusion of this foreign material containing the seeds of life, erupted in protest creating the Moon. It was as if she was protecting her sublime character by creating a satellite that would shine down on the abomination that had disturbed her calm.

Yet, in her reluctance to accept the duality of opposing forces she succeeded only in creating more movement for she set herself up as one end of the opposites and thus began the flow needed for the survival of the new seed. She changed the stagnant environment into one of ebb and flow creating wind currents, flow, moving water - all contributing to the mixing of seed with the environment.

And the Moon provided light in the darkness where there was none before. Thus, the Moon, intended to be a pure expression of feminine, began to interact with the Sun, a masculine presence and in her act of defiance, she only contributed to the further growth of opposites.

Now the environment was set. Growth was allowed to flourish and it did. From the pools of water came life, growing in the shallows, spreading to the land, taking off into the air. And with it came changes in the environment. Environment and life interacted in a great dance to bring harmony to the planet. As life burgeoned, the life itself became the planet and the planet became that life. And the diversity of life sprouted to become like a flower budding and opening to itself as true messengers of evolution.

Truly the pattern had created something of significance. For in the shadows of growth could be seen a potential for great consciousness. And it was here that choice entered into the picture. For with the opposites, there is always choice. For within the two there is the potential of three.

Something wonderful began burgeoning from the evolution of life that was beginning on Earth. There were indications of some new potential. It was in the interactions between the species and the environment.

There was a connection established through the dark side to the enormous movement in the consciousness of the universe. For reflected in the dark unconscious was the potential for the emerging consciousness of evolution. This was unexpected and was attributed to the flow of the environment of Earth and it's place in the dynamic movement of space.

Yes, we desired a place on the edge of the galaxy but had no idea of this development. It was as if each species could sense the dynamics going on and began to see in the changing environment of the Earth the enactment of the movement. As if the movement was a projection of the inner secrets that moved throughout the great unconscious of the planet and her life.

So, there became a new phenomenon as the inner movement was reflected in the outer reality that each saw, and this outer reality created a heightened awareness of what was going on inside. The result was an increase in the rate of evolution far beyond what was expected. This is not uncommon for the intricacies of the pattern of life are so complex giving rise to probabilities that were never anticipated.

Thus, evolution proceeded at a very rapid pace. Animals, flying creatures, insects, fish all flourished, each seeing their own evolution reflected in the outside world, each moving in tune with the cosmic swirls that were and are very turbulent at this place in space. And the diversity became intense so that the dynamics among species added to the evolutionary fuel.

All was going well until that one eventful twist in evolution's path - the advent of man. For in this evolvement was not only the greatest potential for consciousness but also the greatest ability for choice.

The advent of man was a consequence of a number of factors. The environment was such that it supported a creature who could be mobile and make choices. The diversity of nature allowed this. Another factor was a great upsurge in the creative impulse coming from the center of the universe. It is well known that such pulses can rush through the entire universe in the blinking of a human eye and create swirls and currents that have major effect. Such was the case at this time.

The pulse reacted at the juncture of forces between the two stars of Sirius, as you call the Dog Star, and sent a pulse directly to the planet seeded many moons ago by the comet. This pulse brought with it a recollection among some ape-men of their star heritage and awakened some knowing about their own being. This brought a leap of awareness and increased the capacity for inner knowing to a point where additional choices were presented.

This is not very clear. Perhaps it would be better to say that the pulse activated an unused portion of the mind creating additional connections which allowed a new view of the outer reality. In any case, however you wish to express this phenomenon, the group of ape-men affected began to see that they had choices and it was this evolutionary leap that allowed a new species to emerge - the future homo-sapiens.

The word "sapiens" is a key here. For in the ability to recognize choice there was a great inflation of that part of the being that is tuned to the outer world and the safety of the body within it. That is, an inflation of the ego. Big leaps in evolution do not come without some problems.

The difference between the human species and the rest of life was dramatic and it couldn't help but be noticed by the species themselves. They were aware of the big differences and this created an inflated view of themselves, almost giving them a sense that they were heavenly beings, star beings, god-like.

Of course, as with all life on the planet, there was a semblance of truth about being star sourced, yet not in the sense of being king of the hill, so to speak. And because the species gained so much power in its position in the outer world, that connection to the inner realm where the flow of the cosmos was felt began to atrophy because of a lack of use.

There were moments around the fire when the species turned to the heavens and wondered about their origins, but because of survival in a hostile world - hostile in the sense of bodily harm from predators- there was more focus on the outer world and a great imperative for safety. And with the ability to choose the way one lives, this imperative became very creative.

It was this conflict between what seemed best for survival and the wonders around the campfire that propelled the species forward in great leaps. And information about this evolutionary movement flowed back through the cosmic web causing the eye of the Center to take notice.

2

The Center began to notice a shift in evolution. The increase in consciousness was duly noted yet the severing of communication was troublesome. How can evolution be served by such a consciousness if the ties to home are severed?

Drastic measures were considered including the eradication of the species as an experiment gone bad, yet the potential was so great. The pattern in its own way moves like this, weighing probabilities, finally choosing a course of action that seems to account for all the variables involved.

A pulse was sent to the Source that reverberated in the center of the dynamic forces awakening the channel again. More penetrations of the species were tried. This only accelerated the inflation since it convinced some that they and the gods shared a common bond. Rather than helping the evolutionary process, it hindered it further.

A search was conducted to find some beings who had retained their bond to the Source. A small group was located on an isolated piece of land off the coast of what is now Turkey on an island in the Aegean Sea. Their relationship with other tribes was minimal. It was here that further communication was encouraged.

Here the result was more positive. Messages were sent to them in the form of dreams. The dreams were received but the message was unclear to them. Such is the case when important messages are sent to those who have lost touch with their roots.

Fortunately, one woman began to see that there was a call to travel toward the West. She had very little support in this. She was watched very closely by her tribe, yet she was strong and resourceful. At night, she secretly gathered supplies in a primitive canoe and when the currents were right, she set off by herself and allowed the currents and winds to take her. Her bravery was remarkable and showed that the message had selected one who would infuse new life into the lost species.

The journey was perilous taking her out beyond the entrance to the ocean. Her food was almost depleted and her thoughts were confused. Hope was almost gone. And in her delirium of despair, she was shoved by a great wave onto the beach of an island.

When the Sun finally woke her from her weakened state, she was cared for with warmth and given renewed spirit through a dream at which she marveled. It was a dream that foretold of a great new species which she would create. As she foraged for food, she thought of these mysterious dreams.

That night as she crawled into her shelter, with the food and water she had found, she recommitted to the path that she had followed against all odds and settled in to wait. Her wait was not to be long for the pattern was forming around her even then, moving to a crescendo that would be talked about for ages.

The next morning she awoke and went about preparing a small portion of land to help sustain her and keep her safe. As she busied herself with her chores, she was struck down by a violent force. When she recollected this later, the only thing that came to her was that of a small whirlwind. It was not what she had seen, but it was sufficient to give her a sense of reality. For at the moment she was struck down, it was the time when the Dog Star rose from below the horizon and the light from the star traveled into her. Just as the comet had brought life to the planet, so the light carried with it a seed that mingled with her own moisture and she was given life - a life that would begin again the dynamics of opposites.

And so she found herself pregnant with child. She knew that this was out of the ordinary for she had been at sea longer than two moons.

She cared for her child - a son. And as he grew, he helped his mother with establishing more of a permanent settlement where food and water were no longer a concern. The hard days were over and they could relax and enjoy the fruits of the land. And they kept up with their dreams, learning more about who they were and how they were to live. No other humans bothered them, so distractions were kept to a minimum and their evolution proceeded quickly along a path where consciousness and the great Mystery walked hand in hand.

The woman and son lived comfortably knowing that they now participated in the cosmic flow. They were told this by their dreams, although they could not see yet the whole drama they were involved in nor could they consciously know the path of evolution they were on.

When the boy approached the age of manhood, their dreams indicated a voyage to the East was required. So, without question, they packed what provisions they would need for a short journey, for this was the indication they received.

On a calm day under a bright Sun they headed out being

reminded to remember the direction they would travel. This they did by remembering the position where the Sun was upon the horizon when it rose. After journeying for a week being helped by the Easterly winds, they arrived on a strange shore.

It was a rocky place where landing was at first difficult, so they traveled slightly North to a place that provided some shelter from the seas should they turn rough. They made camp and that night they both received dreams about procreation. On sharing their dreams, they correctly concluded that they were to search for suitable partners who could insure their continuation. A caution was given to choose partners that appealed to their sensibilities, having in them the potential and predisposition for inner work.

They struck off heading East after securing their canoe. It was not long before they came across a group of tribes people who looked somewhat friendly. They were living in round huts in a clearing not more than three hours from the shore.

Approaching cautiously, they spied a hut somewhat apart from the rest. An old man and his daughter seemed to occupy this abode. When they made themselves visible to the couple, they were welcomed without any undue caution, as if they were expected.

Establishing a primitive communication with hand signals, they indicated where they had come from and noticed a surprised look on the old man's face. He responded in an excited way that he had been given a dream of their coming and was further told that a journey would ensue. This made the welcome pleasant for both since the woman and young man could see the importance the two placed in the inner way and the old man and daughter seemed to have no hold to this land as they indicated an unwelcome relationship to their fellow tribesmen.

A pleasant visit began, each getting to know the other. The woman saw great wisdom in the old man and the young man was getting on quite well with the daughter who enjoyed an easy grace and seemed approaching a maturity full of wonder and great belief in her father's way of life.

They stayed one moon's time agreeing after that passage of time to journey together back to the woman's and young man's homeland on their island. The description of their place that the woman was able to communicate seemed to appeal to both and their relationships were getting on quite well.

On a day when the wind had changed, they all headed out early in the morning before sunrise without waking any of the tribe. They proceeded back to the canoe and headed out together to the island in the West. Such was the beginnings of the first inhabitants of Atlantis.

The journey back to Atlantis was uneventful and the new family began to set up quarters for their new arrangement. There was an unspoken knowledge that a marriage was taking place between the woman and the old man and between the young man and the old man's daughter. So, they proceeded along this course, each building suitable huts for their privacy in the style of the old man's hut in the land he came from.

During the building process, the newlyweds learned more about each other and began to find a common way to communicate adopting a lot of the language of the old man since it seemed to be easier to learn than the strange tongue of the woman. In the new language, the woman was called Mariam and her son, Allyn. The old man's name was Luthor and his daughter's Morgana. And so began a period of conjugal bliss.

Morgana, who they learned was not actually a biological

offspring of the old man but a waif who was found abandoned in the forest, suited Allyn beautifully. They both eked out a magical quality that surrounded all that they did together. It was as if their inner most souls were in constant conversation and their delight was evident on their faces.

The relationship of Luthor and Mariam was based on mutual respect and honesty. They shared together much of what they believed about life and the attraction of their souls had a maturity that gave grace and a presence exuding almost a royal quality. Life went on in this way and each was graced with offspring who brought with them the happiness of children to the island community.

All went without incident until one day while Allyn was hunting in the woods, a strange occurrence happened. While sitting still awaiting the arrival of game that normally passed this part of the woods, Allyn began to see an image in the shadows of the tree limbs. It was a man in a dark cloak standing there looking at him. Allyn knew this was no ordinary man, but someone from the place where dreams originate.

After he became accustomed to the image, he waved a greeting. The cloaked one approached and began to speak.

"You are our envoy to this planet. We are the Ancestors who created life here. We wish you to know more about your purpose here. It will not be dangerous to you. All will be imparted to you in a way that does not threaten you and cause you to do anything you do not want to do. Is this clear?"

Allyn nodded in agreement.

"We will meet like this for many moons while we give you these lessons. Do not be afraid. All will be well." And with this, the image faded from view.

Allyn, startled at this meeting, rushed back to his family

without any game to tell them what had happened. His mother was not surprised since only she knew of his strange beginnings. She had not shared this with anyone. Yet, that night after they had encouraged Allyn to continue going to the woods each day to meet with the cloaked one, Mariam related the story of his conception not knowing how to explain it to Luthor.

He, in his wisdom, had already suspected his specialness and told of his own belief that his adopted daughter was left by the Otherworld for him to find and raise. He was very supportive of his wife's story, helping her to assuage her doubts and concerns of what happened for he was firmly convinced that many things happen in the world which cannot be explained in human terms. And so he promised he would help her son bring the messages he received to understanding in any way he could. This he pledged to his wife.

So, there arose a period of time when many messages were passed from the cloaked one to the son, and the understanding was achieved through the help of the old man's wisdom. The messages talked about a way of living which was far different than the family had heard before in either of their homes. It was a way of living based on an inner way that went beyond their understanding of how life was.

The messages talked of the nature of human beings as made up of the planet's body and also the energy of a far away star. Allyn was told about the type of energy containing opposites: dark and light, male and female, consciousness and unconsciousness. He was told that the unconscious energy was mysterious and unending. It was here that the flow of the universe was felt and that feelings were the surface of this place, like the waves on the ocean. It was in these depths that humans would find the answers about who they were and why they were here on this planet.

It was the task of humans to delve into this dark place and uncover the truths that needed to be revealed about themselves and the moving firmament that made up the place of the stars and planets. They were instructed to listen to the messages of the dreams and learn the language where objects and actions in the dreams were symbolic messages of who they were and what they needed to learn.

They were taught the language of dreams that comes from the timeless realm and after a time Luthor began to see how this language contained much power. He recalled stories from his own father that told him how even the events in the outside world could be seen through the veil of this language so that the distinction between the outer and inner worlds became blurred and much inner knowing could be achieved.

All of the family seemed to take to this learning very naturally and they began a ritual of telling their dreams to each other in the morning to determine if some new direction was called for in the group. They also assisted each other in seeing the personal messages contained in the dreams which helped them learn more about themselves. The children were allowed to take part in this activity as their knowledge of dreams increased. In this way, the family group became very close.

Allyn continued to have meetings with the cloaked one and the messages helped them to work with their dreams, instructing them on the process with which they were working. There was no fear of this dark place within them as Mariam and Luthor had both experienced in their homelands. They rejoiced in this freer life and wished they had this knowledge before and perhaps they would not have been persecuted for what they knew was right and just.

The family enjoyed their new found way of inner knowing. Allyn continued with his sessions with the cloaked one. There was much imparted to him in words, but more just in the presence of the Ancestors. He was gaining much confidence in the manner of this way and was taking more of an active role in the dream sessions they would have as a family in the morning.

Allyn's insightfulness at understanding the dreams of others began to cause friction. As each began to open to changing, more of the insight was given as to how this change could best be affected. This required going further within and Allyn could see the messages very clearly, yet there was a reluctance among some to go that far into their inner depths. Thus, his sharing of messages or his help with the meaning of the dreams at times was not appreciated.

He noticed this, so he backed off not wanting to hurt his family members. But then the cloaked one chastised him for this and Allyn was caught between doing what he was being taught and seeing the distress among his family. The cloaked one began to show him more of what this all meant. And it was this that he brought back to his family to try and show them the seriousness of the work.

He told them that the conscious part of the human had a tendency to want to be in control and be independent. This was good for survival and contributed to the growth of the species. Yet, he said, the danger was that the conscious part could grow further and further away from the inner Mystery which would sever the bond to the cosmic flow. This evolution of the species would not be in harmony with the universal movement and the species may suffer great destruction because of it.

He went on to teach in his quiet way the need to

relinquish control and be willing to join with all the images of people and other living creatures that appeared in their dreams. He knew this was difficult because it required great change. He told them the change was overall a movement from independence to a group sharing or intimate relationship with their inner portions who wished to bring the conscious part into the fold. He said it was allowing the conscious part to be fully in communion, at first, with all the faces of the inner Self, the center being, but then also being a full participant in the cosmic flow. This journey from independent consciousness to a full interrelated way of being was a difficult challenge but one which would serve the human species well and would in turn assist the evolution of the All.

His message was received thoughtfully and they all realized their reluctance to participate fully. Yes, they were all called to the inner way, but there was also fear of seeing all that was being asked. But they received the message as truth and marveled at Allyn's understanding of what was, for them, a new language - at first difficult to learn and then frightening in its global perspective. Yet, they were all willing to follow the messages of the cloaked one as related by Allyn.

And so they began more of an in depth view of their dreams and began to go deeper into themselves. This helped the children as well and as they grew in years, they could see the tremendous changes that were affecting everyone in the family. And they were at an age where change wasn't such a high wall to get over.

The family went on this way, learning more about what their inner life was saying. At one session, Luthor was given a dream that seemed to tell him of his biggest transition - death. This was inevitable but was taken hard by all who had grown to love him as the benevolent head of the family. But much was learned about this first death on the island called Atlantis, for much was promised about new life as they saw the messages Luthor received.

As for his part, he was ready to depart this life and find out more about the great scheme of life and how it all related in the world to which he was about to travel. So, his excitement at this prospect made his death more of a celebration of sorts and relieved the mourning, balancing it with the hopes of the old man.

Luthor's last message given to his family was one that shocked many except Allyn, for he had already contemplated the need to visit the old man's homeland once again to recruit other members to their fold. But it was better coming from the mouth of one who was respected for wisdom.

And so after the death of Luthor, the family talked about what was needed and how to go about accomplishing it. The journey to the old man's country caused many misgivings in their hearts and it was fair to say that their enthusiasm for the trip wasn't quite enough to propel them with love much as it was with the practical need.

3

Before the party left to find new recruits in the homeland of Luthor, Allyn had a session with the cloaked one who explained the apprehension within the family about this journey. Much had occurred since the time they had brought Luthor and Morgana to Atlantis. There were many wars among the tribes and there was much more focus on survival than on spiritual matters.

Those who had a leaning toward the inner way were rewarded if they could anticipate the outer movement in an advantageous manner. This was the time the reading of entrails of animals began and much work was done to apply what they had remembered from the ancient times about what external things meant in a symbolic way. Those who followed the inner way for personal growth were a dying breed.

This was all to explain to Allyn that it would be difficult to recruit followers of their spiritual ways in the same manner that Mariam and he had done before. New drastic measures would be needed. The cloaked one said they would have to recruit children.

This shocked Allyn. He had no stomach for such drastic means until the cloaked one pointed out the difficult life most children had in this time of war. Many were orphans barely scratching a sustenance from the earth.

After much soul searching and talking with the cloaked one, Allyn agreed that it seemed the only way to keep his family growing with the diversity they would need for a healthy population. He related the story to the family and there was much discussion and some anger at having to do this but in the end all could see the wisdom of the cloaked one.

And so they selected Morgana and Allyn to go with some of the more robust children who could understand the mission and they set out on the Easterly wind. Following the same route as before, Allyn navigated them to a protected part of the shore and they set up camp. The journey had been about the same distance as last time, a short two weeks in duration.

The land was indeed changed. Remains of battles were ever present and the smell of smoke from recent fires was in the air. They decided to head slightly North along the shore until the smoke cleared hoping that they could avoid any marauding bands of warriors. It was not to be the case and many times they had to run for their lives as they were spotted and thought to be invaders of the land. But they managed with fleet of foot to outrun the weapon-laden men and women and sought shelter in the dense woods where hiding seemed to offer the best alternative.

They traveled this way hoping to come across villages where there might be many orphans who could be cajoled into taking a long journey. They had to travel inland further than they wanted but eventually came to a village that seemed to have recently suffered a devastating attack and there was much activity to set things back in order.

They approached cautiously and tried to stay on the outskirts of the village but they were quickly spotted and were mistaken for locals on the run from the attacking hoards. They fitted into their roles and shared with the villagers their narrow escape from a fierce band of men and women. As Allyn began to talk, the villagers were mesmerized by his pleasant speech which captured their imagination and they were quickly welcomed and given food and a place to stay.

And so they had time to look over the area and see how the plucking might satisfy their need without causing too much grief at their departure. After seeing the devastation of families after recent attacks, they could reconcile better this action they were about to take for there were many homeless children who longed to leave this country of war.

As the children of Allyn and Morgana helped in befriending many children, they were invited to join the newcomers in an adventure. The adults of the village appreciated the attention the new travelers gave to the orphans and other children and encouraged it further as additional food was brought to them to help feed the collection of children they were gathering.

The trust the villagers had in these new people was amazing to Allyn but Morgana could see the respect they had for him and how they admired his ability to speak and tell stories. In Morgana's eye, Allyn was growing into his wisdom of the inner way. She could see how he responded in a manner that touched the people's souls in ways they couldn't see but could feel something wonderful happen to them. Morgana warned Allyn that if he kept up this way he'd have to take the whole village with them and this was not possible.

So, Allyn, unaware of his growing talents, took the advice of Morgana and began to pull away saying the children needed his help. They moved their little group to the outskirts of the village and one moonless night they moved out in a group to bring the children to their new home.

The route back to the canoe took longer than they expected because of the tribes that were moving to kill other tribes that seemed unfriendly. It was a very chaotic scene, one that scared the children and changed any thoughts of their staying in this place of war.

They finally reached their canoe and waiting for the change of the winds they talked to the children about their new home and how pleasant it was. They understood that they would be homesick for a while, but all would be good soon. As Allyn talked, Morgana could see words enter each child and how the words comforted them.

She loved her husband so much at these times. Her thoughts returned to the time when they first met recalling her inner feelings about her own recruitment and so she was able to give comfort to the ones who were still quite fearful. The winds changed and they began the voyage back to Atlantis.

The return trip was without incident and the children were taken by Mariam to get settled in. She had worried some because of their late arrival but her dreams indicated there was nothing to worry about so what could have been a painful wait was not as bad as anticipated. The children of her own marriage with Luthor and the children who remained behind from Allyn and Morgana's marriage were sufficient distraction to keep her thoughts on other things.

She had carried on their dream sessions and told Allyn on his return that their recent young boy showed dramatic signs of understanding the inner way far beyond his years. She confided to him that she saw the spark of the Dog Star very prominent within him. Allyn took his mother's word and began to watch the young one for similar signs. However, it was the confirmation of the cloaked one that finally convinced him that he had an heir.

And so began a period of building. Mariam suggested they move camp to a more suitable piece of ground they had seen recently, closer to the main mountain of the island. Here appropriate space was available for future expansion as the population of the island grew. They laid out the new place in a circular arrangement to remind all that the journey of humans always led to the center place where they could live more in the flow of cosmic evolution. They built huts for the children and cajoled some of the older children to be house mothers although they didn't appreciate their new title. But they did begin the education of their new inhabitants in a place where they could understand. In this way, they were introduced to the teachings of the cloaked one as relayed by Allyn. Most were interested, but a few felt bored with their new inner studies, not seeing the broader picture. This was to be expected and nothing was thought about it, yet seeds that would destroy the tranquil Atlantis were even then present.

As time went on, Allyn became more attuned to the messages of the cloaked one and he began to understand who they were. While the image he saw never varied that much, he understood that he was being guided by a group of spirits who had purview over the evolution of the human species and other inhabitants of the planet Earth. They embodied the essence of the energy of the Dog Star and were quite powerful. He knew this from experience as they increased their energy surrounding him as the sessions progressed and he could feel the potential of their great presence.

He began to refer to these spirits as the Guardians for they did oversee the evolutionary movement of the species providing guidance when necessary to keep the growth as close to the evolution of the cosmos as possible. He also began to understand the movement in the universe as like an ever-changing spider web that connected all living things in a network of love. Because choice was essential to the nature of humans, they could be part of this web or not depending upon how they acted. Independence and the growth of consciousness to serve only the mind of man contributed to the severing of these cosmic connections.

He was constantly warned by the Guardians that the conscious pursuit of an individual's goal independent of the cosmic flow was tantamount to ultimate death. The Guardians explained this in a vivid picture that Allyn related to all. It was of a little boy pushing through a spider web to retrieve his toy. In this action, all could see the pursuit of a human's wants and desires independent of the web to be detrimental to the whole.

And so as he understood more, he taught more to his people. While all were respectful, there was some dissension by those who could not understand fully for they had not reached the center place where the love of the All was felt the greatest. It was not their fault, for seeds of attaining worldly power were planted in them during the age of war they had lived through. The resistance was minor though and all persisted in listening to their dreams and continued to see where this journey would lead.

The cloaked one did suggest some external symbol might be valuable and led Allyn to a black stone that he was to place at the center of the village as a symbol of the center where a solid foundation in cosmic love could be achieved. And so one summer afternoon, some of the new adults helped Allyn place the stone and the village helped him commemorate its position in the lives of the Atlanteans.

Growth of the village congregated around the center stone

as time went on. Age crept upon Allyn at the death of Mariam. She had always been the matriarch of the village and now the mantle was passed to him and Morgana.

It was a sad time for them all at Mariam's passing. They never found a trace of her after she left her hut one night to die. Most thought she had been eaten by the wild animals, yet others thought she had walked off into the sea for they never found any bones. This was not to be the last mysterious death for those who had the essence of the Dog Star within them, but at the time it was not understood but explained away so that peace could reign in the mourners' hearts.

The passing years brought growth to the village and an urge by some to build in a more permanent style. And so, stone homes were begun as better insulation from the ocean breezes that could chill even the hardiest person to their core. And irrigation was established for their growing crops that were planted. All the new building was faithful to the circle as it was the essence of who they were.

As years went on, Allyn was seen more at a distance as he seemed to be troubled with all the new building. Perhaps he was already beginning to see the demise of his world as the interest in man's creations began to take precedence over spiritual matters. It was not a sudden change but one that gradually came and only one who was attuned to the heart could see it for what it was. Morgana could feel it also and suggested perhaps more recruitment but the cloaked one was reluctant to expose any of the Atlanteans to the further influence of Luthor's homeland.

Allyn's teachings began to appeal to a smaller group, those who seemed to be more attuned to the spiritual life and this pleased Allyn greatly for he could see the growth of these people. It was gratifying and balanced his inner feelings that foretold of a tumultuous end to this way of life. He taught them about the nature of the cosmic web and the pattern that manifested evolution. He taught about the All as a living being made up of all living things in a living web of energy. This led to discussions of what it meant to be a member of the inner group that was within a human, and also to be a member of the whole for the nature of the universe was "group essence."

Many could see how these teachings would not be received by those who served the ultimate authority of man. It was a great contradiction that as man evolved to fill his own needs, he desired to be in charge of many people and this gave him power which would never satisfy. Yet, as one gave up the prominent position of the conscious self to meet the needs of the inner group, he gained a feeling of fullness and a sense of purpose that was long lasting and full of the power of the All. Stated simply, as the conscious part of humans gave up his or her independent position, he or she gained everlasting purpose and meaning, while in maintaining independence they gained power that was short lived and unsatisfying.

Yet, the plunge of the conscious part into the inner way was fraught with difficulties and it was a hero's journey for the bitter truth of who each one was had to be faced. But, in the struggle, each found their mission in the grand scheme of evolution and, according to the Guardians, the scheme of evolution was served in the knowledge of consciousness. These lessons were difficult for the small group who gathered around Allyn, but laid the groundwork for the ultimate day when Atlantis was no more.

This small spiritual group became Allyn's greatest support. Morgana participated fully in the happenings of this group and she came into her own, so to speak, as she led ceremonies to commemorate various high points in their spiritual life. It was this group that was the beginning of the spiritual leaders of Atlantis and many wonderful experiences were related about the goings on of this group, for in following the inner way many miraculous events can happen and do.

The formation of this group was not without a purpose in an external way for the schism that was beginning to be felt as the population increased required much spiritual tending. Families that had lost the conjugal love provided by the cosmic center were split in their ways, part going toward the material and part going to the great Mystery that was within all humans. And so much healing was needed.

The group combined their knowledge of the inner way with external healing methods using herbs and other plant material as they were directed by their dreams and visions. Prior to this, physical sickness was very rare as most healed their spiritual bodies before the illness manifested into the physical. And even when the sicknesses increased, it was always the way to try and heal the spiritual body first, helping the person to come to grips with the holes in their personal web where a lack of understanding of who they were and how their actions were at odds with their inner direction was promoting the illness.

Yet, out of compassion when these methods were either rejected or too slow to heal a physical illness, they resorted to remedies that worked or didn't depending upon the disposition of the sick person. No judgements were made for the spiritual group had long ago realized that this path of schism was the inevitable conclusion of Atlantis and could see it coming long before it became known to the general populace.

With the people leaning more to the spiritual group for their physical well being, they desired to place them in a higher place than themselves and so without consultation with Allyn, they built a temple to signify the importance they felt in their spiritual well being. The temple was built on a hillside near the natural harbor of the island and became the focal point for all activities.

Allyn was much distressed at the building but it was beautifully done with much love crafted into the work so he relented and took up residence along with his followers. Morgana became the spiritual leader of the temple performing ceremonies with her followers as the needs arose. Allyn was to be known as the recluse that he was, but all knew his wisdom was imparted to the spiritual leaders who ministered to the people.

As Allyn approached his later years, he spent more time with the Guardians, learning and imparting their wisdom to his people through the spiritual leaders of the temple. As Allyn walked around the temple and viewed the growing city outside and its many activities centered around material matters, he wondered how they could have prevented this, but he also knew that further evolution would be required before humans were ready for the next leap in their spiritual journey.

And so, like the alien that he was, he kept in the background and conversed only with those who tended the flock. His greatest joy was being with his son, who gleamed with the Dog Star light and had gained as much insight as was possible given the state of the island. He assisted Morgana with the ceremonies and became very popular with the people for he had a gift of the tongue like his father. He could mesmerize the throngs who attended the temple on special days with his speeches about the potential grandeur of the human species as they assisted in the great movement in the cosmos.

The activity of Atlantis continued with much building including a fleet of ships to trade with neighboring states for needed staples and niceties that were desired by the people. And so, the evolution of Atlantis was cast in stone and the inevitable was close at hand.

Years passed and more building ensued. The temple was expanded and a larger black stone was placed in its center and carved in the shape of a large rectangular block. Here the center of the worship was held and the stone took on a purely physical significance rather than the symbolic essence of the inner center foundation.

Allyn withdrew into himself more and grieved for the journey of his people. He wracked his inner knowing for answers to this turn of events that brought a feeling of tumult and dissension in the island nation. Now Princes vied for the rights of trade routes and began to accumulate gold as a sign of power. Allyn could only grieve as they lost their own inner gold.

Frustrated, he turned to the Guardians who had been quiet for a time to let Allyn grieve. Finally they returned to face Allyn's question - why had he been sent here as an envoy when all that seemed good had faded into a world that focused only on the external. He was angry when he asked and was willing to take the brunt of their criticism of how he had handled the movement of his people. But, the criticism didn't come. Only praise. For he had not realized the extent to which he had been an instrument in the evolutionary movement.

The Guardians talked to him personally now, showing him that his mission had been fulfilled in a fine manner. For he was not here to save Atlantis, but to be the instrument in getting the message to the world. The lessons that Allyn taught were still present and marveled at by the people, even though they had turned to an easier way of spirituality. Such was the case at this stage. He was told that was inevitable. It was only the message that needed to be seeded here. And so it was.

They showed him the message being even now taken to other lands as the ship's captains bragged about their spiritual leaders and shared with other inhabitants the importance of the circle and the golden center, and the path of enlightenment that existed in the inner darkness and how change was essential to all that existed. Allyn could not believe these visions at first because he did not know the great import of his words. He saw how the words relayed to others struck deep into some and began to immortalize the place of Atlantis as a mystical place where great inner knowing went beyond the normal. It became a magical place for them which they had no desire to visit but still marveled at from a distance.

Many spiritual ways that were related circulated through the land of Luthor and were adopted by the tribes. And stories were told which contained much symbolism of an island of magical beings who came to the Earth to bring the land of the gods to earthly beings. The stories were distorted to fit the beliefs that were held, but were potent in the meaning the local inhabitants gave them.

And Allyn was shown visions of the word spreading around the world as people carried the message. In a land of desert sands, many rituals were held to honor the Dog Star and myths were created about how the opposites could not be held in the egg of the universe as man's way persisted over them. Such were the stories perpetuated about the struggle of man against the Great Mystery and the consequence of man's ways as he lost the guidance of the deep.

Allyn marveled at the changes the stories were given and wondered how such alterations could move the human species to their eventual evolution leading to something more. The Guardians commented not about this, only pointing out the power of the symbols that would work their movement from within until the day the species could see the ultimate death that they faced with the ways of man. Allyn could only have trust that the lessons he had brought to his people from his spirit guides affected the right push needed for evolution.

He thanked the Guardians for the visions and was now content knowing he had not failed but had begun a movement that may eventually move the human species back to a way that was in harmony with the cosmos. So, he knew his work was done. His end was at hand and he prepared himself for death.

He could see that the inner group of the temple kept the lessons to heart and could maintain them in spite of the dominance of the external activity and focus. Morgana had passed on her great feminine spirituality to select women who vowed to protect the lessons as time moved on. Allyn's son had heirs of his own now, and while they consorted more with the Princes of wealth, they contained the seed within them as they moved throughout the external world.

One morning on a brightly lit day, Morgana and Allyn sat in their rooms contemplating it all. She fondly remembered her own beginnings in the forest of old and how the spirits had cared for her waiting the arrival of Luthor. She longed to return to the forest now and become one with the forest spirits - a place of home for her. And so they planned a secret excursion to Luthor's homeland.

Helped by their close followers, who understood such things, a canoe was prepared for the two lovers and they left under the escort of two trusted sailors. The journey was without incident and they beached on the shores of what was now called Britain. Here Morgana followed her developed instincts and led them to a dense forest. At the center of the forest, a lake surrounded an island that only Morgana could see. Here she turned to her husband and said her tearful goodbyes as the two sailors watched in wonder and amazement as a small canoe floated out of the mists.

Morgana, with a sweet smile on her face, said her last goodbye to Allyn and climbed aboard. She was never seen again although her legend was spread through the land as the Queen of the Fairies. Such was the way of the times. Allyn returned to Atlantis with his crew and settled down to contemplate his own leaving of this life.

Allyn began to realize that he knew nothing about dying and didn't know where to go with his own preparation. So, in desperation, he sought out the Guardians. We were not surprised when he came, of course, because what he was about to undertake was not only new to him but also new to the planet. For within Allyn the source of the Dog Star was great and was not to be recycled in the same manner that Earthly souls do.

Allyn came to us on a bright morning in the forest where we had originally met. He was nervous about this meeting for he had no idea what was ahead of him. We appeared when he called and quickly set him to rest by saying his end was not to be known up to this point and he shouldn't feel lost because he was not able to deal with his own departure as had Morgana, for this was the reason he felt weak and inadequate in addressing his own demise. Once he recognized this, he was in a much better mind to listen to what we had to say.

We told him again of his miraculous birth, reminding him he had not been fathered, but sourced from the Sirius system to be instrumental in evolution. As such, his departure from this time in the Earth's history would be something other than the deaths he had witnessed, for his spiritual body would be made incarnate again at a later time. We then explained that the timeless realm that we came from was able to see this future time and we would allow Allyn to join us in our wait. He, of course, did not realize at the time that a place had been reserved for him in the spirit world at the center of our circle.

We told him that he would now be initiated into the circle of our energy and asked him to follow us. He was quite distracted by now, but willingly followed us as we led him to a nearby cave that even he knew nothing about. The cave's entrance was clouded with a veil that was a doorway between worlds. It was through this veil that we led Allyn. His nature was such that all Earthly manifestations were cleansed from him and he stood in the cave in his spiritual body. He was not aware of this.

The cave was deep and we led him into the darkness of the depths. As he became accustomed to the darkness, he could see us surrounding him for he saw us as we truly are. We cannot explain this in human terms, but Allyn was not frightened of us and accepted his place in the center.

We carried him for many light years to his home with us and he stayed here until he was needed again. It took him awhile to get accustomed to this state, but eventually he became quite content being in our center. The length of time he was here can also not be explained in human terms, for his sense of time changed when he went into the cave.

Back on Atlantis, his presence was sorely missed and

many legends were spread through the world at that time about his lessons and who he was. Many marveled that such a being came to Earth and they gave him many magical qualities that Allyn would have been surprised at, for to Allyn he was simply a messenger on Earth doing his mission.

Atlantis went on to her demise and the survivors were sent by the old holy man to the European coast including the Northern coast of Africa, and to the center land connecting the Americas.

Among the travelers to Europe were those that were hybrids, carrying a portion of the Dog Star seed. These became those who carried the yearning for the inner way and became priests in Egypt and later the true alchemists who sought the inner path. Many secret organizations were formed around them.

Those who traveled to the New World carried the unique feminine energy of Morgana who had vowed to keep the lessons. They mingled with the New World natives and carried the Great Mystery throughout many centuries creating many myths about the next evolution, notably the myths of Quetzalcoatl and Tezcatlipoca.

Religions came and went all seeking pieces of the story and many symbols persisted that spoke of the great seeding on Atlantis. Some still seek this place but they will not find it until they find their own place in the cosmic flow.

4

Allan put down his pen and looked at the paper that was in his hand and the words that described this fantastic event in the past when cosmic forces were alert and available to help humans climb out of their sleep. He knew in some part of himself that the story was true, but in the part of him that was aware of the precepts of society, he had a difficult time owning what was not acceptable - a story from long ago that seemed fantastic, outrageous not in the telling of it, but that it could involve him.

That was the rub. If a story like this had been encountered in a book, he would find it interesting and fascinating. He would see the possibility of it, marvel at the action that was undertaken by the spirit world and mourn the fact that it didn't take. Then he would realize that it left a legacy for the human species. All this would go through his mind in an instant, only to be erased by the thought that he was there.

This was what scared him for it brought out his inner movement that resonated with the story and he realized that the story was not fiction but an actual occurrence in which he was involved. This bordered on the insane for if he accepted what he was feeling, he would put himself on the edge of society where he would be seen as someone who was delusional, demented or some other scandalous category given by a society that did not accept such stories as truth.

He got up and walked outside, wondering what was happening to him. He saw nature in its full power - the wind moving the tops of the trees, and the noises of insects and birds scurrying around trying to find the food that would sustain them for the day. The smell of mold and earth and the moistness that promoted growth was around him.

He relished this place in the woods, yet it had brought him to a place that was uncomfortable. He was toying with his very beginnings in a way that tormented his understanding of this world - the world he had grown up in where reality was the physical, the material and anything beyond this view smacked of insanity.

He walked through the woods for comfort for this world of life did not reject the miraculous. It moved him to a place that allowed the surfacing of his feelings, knowing that part of him which saw beyond the world of reason, yet at the same time uncomfortable with his unique being that could see the mythic proportions of life.

Many tumultuous eruptions began as his soul struggled to surface, held down by his concern about how he would be seen in this society. He thanked the woods and his cabin for providing the shield that he needed, thankful that he was not exposed in a city where he knew all eyes would fall upon him, as if they could see the crazy thoughts that were moving through him.

He remembered his fear as a child being near a place for the insane - the funny farm. A place where parents told children they would go if they didn't shape up. He remembered driving past the stone walls of this prison/hospital where people talked to themselves out loud, and screamed in the night seeing visions that brought fear to their very souls.

His childhood fear of being sent there laid out before him and mingled with the strangeness of his thoughts now. How could he remember these ancient scenes and know he was there? Even more outrageous, how could he identify with the hero of the story who was born part human and part Dog Star energy, whatever that meant? And how could he hear the voices of the Guardians, and know that he had been with them in this place called Atlantis? It was all unreal in the world in which he lived. All too fantastic to admit. Yet, there it was moving in him as if some alien was awakened in the bowels of his being and now wouldn't let him be so he could capture normalcy, so that he could show his face in public again feeling acceptable and productive in his role as one of those who contribute to the world of this age.

He cursed his work and stumbled through the woods finding no solace in this aloneness that he had sought in order to get in touch with what was bothering him. He certainly didn't want this. He wanted to find out who he was, but not this. Who he was meant an occupation, a call in life to add to the society's wealth. Now he was confronted with something so strange that rather than bringing him closer to a society he felt foreign in, he was moved to a place that further alienated him making him fear for his own sanity.

No, he didn't feel crazy. That was the problem. He felt more true to himself than he had ever felt, yet he knew how he would be judged. He thought of family members who would have him committed. He thought of psychiatrists who would love to get hold of him, to bring him back to a place where he could join society again, dispensing pills that would elicit normalcy and shove all these strange thoughts into the darkness again. These thoughts destroyed the view that this society wanted to keep and hold onto before they all succumbed to the call within that would lead them to behavior which was not acceptable.

Acceptable! he thought. He could see the falseness of the beliefs, seeing only the material world. He knew there was more, and he followed a way that exposed the inner rumblings of himself. Why should he be surprised at this outcome? But he had exposed too much and had lost touch with his own sense of belonging.

Yes, he followed his dreams, but he could still face people in a normal way, for it still had the semblance of psychology and while most did not look at dreams, there was a recognition that such work could help someone discover what was bothering them. It was not in the same vein as science that people believed in whole heartily, but at least on the edge of acceptability.

Now he knew he had gone over the line and he had to deal with this and he wasn't looking forward to it for it brought up things that he didn't want to look at. It was as if a beacon was shining on him, even here in the woods with thousands of eyes bearing down on him looking in disdain at this lost creature who now thought he was someone reincarnated as the messenger of ancient lessons.

He rejected this, as he stumbled around the woods aimlessly, lost in torment that had only now erupted changing his place in the world to a place he had no desire to accept.

5

We will begin today with a message to all those who are willing to listen but fear where it all will lead.

We tell you that it's natural to fear this for you are unaccustomed to working with the inner realm. This realm is infinite and can lead you through all the different parts of your whole Self - those parts of yourself you are not familiar with. It is here that your talents lie and it is also where your center Self or true Self resides.

You are born into this world without any knowledge of these things. Unbeknownst to you, you come from a realm that is not like this world, for it is a place of spirit. Thus, your nature is attuned to this spiritual place for you come from there.

Yet, the world of consciousness presents much that is different. There is the world of seeing, hearing, tasting, feeling, smelling - all physical aspects that you had forgotten and it takes much adjustment to attune yourself to this new world into which you are born.

And then there is also the way of your culture. Each culture is different and requires much learning to adapt to their ways. Much that you learn today is associated with this - how to adapt to the culture that man has created for himself.

It is a male dominated culture and so there is much emphasis placed on doing. The art of being who you really are is not stressed. What is really required is for you to learn the place where you are needed to sustain the culture.

Much effort is placed here. It dominates your waking hours and consumes all your interest for it is, at first, all you know. Memory of what came before or of other worlds that exist at the same time have no meaning for you. At times, some enlightened parents can make you aware of other worlds, but this is rare.

So, you begin to find your niche in society and begin the process of living up to the ways the culture presents to you as proper and acceptable. It is the acceptance of this major role in society that contributes to your fear. You have worked hard and long to create a space for yourself, and to now open up to a different inner world that brings messages to you about yourself is a threat to the place you have created because modern society has not made room for the other world.

If you are drawn to the inner way, you have much to deal with. How will you be seen by others if you begin to listen to your dreams? What will become of you if your dreams lead you to a journey that destroys your place in society which you have worked so hard to attain? Where will you garner your self-worth once you begin to live an outer way while you listen to your inner way? Who will have authority over how you govern yourself? And what is the role of your conscious self once you begin to listen to your inner voices?

All these and many more challenges will face you. Is it no wonder that you fear opening the door to a world that makes your current condition seem so shallow on the one hand, yet threatens the way of life that you know and to which you have become accustomed, shallow or not?

Yes, there is much to deal with. There are all the ways you have learned which may or may not be appropriate to who you are. How many times do you feel as though you are acting for the audience of society, constantly adjusting your behavior to appease those watching faces who become critical when you don't conform to the acceptable ways? How many times have you suppressed feelings that tell you what you are doing is not satisfying and fulfilling just to maintain a demeanor that is pleasing to others, especially your loved ones?

It is difficult to accommodate everyone and there are so many rules to remember and abide by, yet here is another world, an inner world which seems to have a whole different dimension with which to deal. It seems so overwhelming to encounter a world that encompasses the All, and you realize that you can barely keep your head above the water in this outer reality that you live in. Yes, there is much that you have to confront.

Yet, are your feelings leading you there? Do you feel empty and without purpose in the outer world of man's creation? Why is that? What is it that motivates you in your life? Why are you here? Is it just to make a living so that you can live in accord with some standard that seems less spiritual and more material?

These are the questions you have to address in the quiet of the night when you are alone. It is the time when you can feel what goes on if you will just allow it. How can you know where you are if you constantly take chemicals to mask what is happening within? Can you get over the first fear and allow the feelings to be with you?

Feelings are the waves on the ocean of the unconscious. They teach you what it is that is calling you. It is an enticement to enter the water. Feel the feelings. It is your first entry way to the world that can help you sort yourself out and distinguish between who you really are and what behavior you have placed on yourself to abide by the rules of society.

Yes, the law of the spiritual realm is different and only wishes you to be whole. Can this be such a great leap for you that you haven't the energy for it? Can you not adjust your life for yourself, foregoing some of the busyness that consumes you?

Begin with your feelings of fear. Answer some of those questions that arise within you. See where that may lead and try to understand why your fear is so great. What is the worst that can happen if you begin to listen to yourself? Death?

We all go through many deaths throughout our life. Many losses. Many disappointments. These are all like small deaths that we experience. How did you deal with these deaths? Remember. See where you are with your feelings and how much you have covered over. You know this.

And there is much fear in feeling these feelings again. You think it is best to ignore them and carry on. Yet, you know you begin to feel like a zombie going through life having no feelings except those just at the surface that bring you down just before you take some chemicals.

You have heaped so much protection over the feelings, to open now seems so threatening to the very sense of your self. Yet, what is that sense but a weak facade that keeps you from knowing what lies within you.

It is up to you, for you have the choice. Address your fears and begin to listen, or carry on the way you have been. We can only say that your true nature lies within and you have the courage deep inside to do this. Both reside inside you.

Where do you go from here?

6

After walking through the woods, Allan was in a better place to address his fears. For all his many years of life, he had tried to please people and be liked by them, but at the same time he felt callings to things that others shied away from. This caused a constant tension to be in him which occasionally erupted into anger at those who were close.

This tension was always with him and contributed to his great capacity to hold this point in his life - the tension between two world views. One view was held by society in general - the great belief in the external physical world as the only world that could be trusted. The other world view was larger, containing the physical and the spiritual, or those aspects of the world that could not be seen per se, but could be felt by senses attuned to this world of subtle living energy. All his life had been preparation for this time.

He did not grow up independent of the society he was born into. He, in fact, became a student of its inner workings as he was led through their schools to learn the inner mechanisms that kept the world humming. And because of the tension of opposite worlds within him, he was able to see the advantages and disadvantages of this way of life. For how could he participate in delivering the message in a way that would speak to his people without knowing how they thought? All this was in preparation for the time when the Guardians would deliver the messages again to the people of Earth.

Allan mused about these things late into the night and fell asleep close to the morning time still in turmoil, not about his preparation, for he could see that with his peripheral vision, but about having to carry the burden of the messenger. He was not prepared for this, nor could he be. It is something that just comes upon a person when they finally reach their center place and the full brunt of their mission in this lifetime comes upon them.

How can one get used to it first when one has not reached the center place where it is kept? There is no warning, for it is only known once one has done the inner work and arrived at the place where such things are known. So, Allan's shock was not unexpected. And he was propelled into the time when the greater picture had to be internalized rather than just viewed from a distance.

As Allan awoke late in the day, he came to realize that the work ahead would not be easy. By now he was astute at reading the inner signs, and he could feel the tension of the two worlds within him wondering if, as he addressed this tension, it would propel him into a new creation that could lead people to a new place of awareness. This leading role suited him as he lay in his bed contemplating where he needed to go from here.

His dreams were muddled, a confusion of action that he had difficulty putting into words, a sign that his energy field was in movement - a movement that would help him comprehend what fears were at work in his very depths.

Allan began the day late, realizing time was passing which

always bothered him. Time was at times a nemesis for him. He prided himself on his quickness and how he could come to conclusions faster than most. Now he had no clue how to resolve his own turmoil. But time beat her steady beat and his awareness of the passage of time seemed to heighten his own awareness of how lost he was.

He tried to sit and wait for inspiration but the time forced his action and he was unable to get to any conclusions. Time at these times becomes an enemy, pointing out the passage of life where no resolution is possible. It is severe in those who are carrying a heavy burden that must be resolved in order to move on, yet the opportunity given by time is not recognized at the moment when the turmoil is heaviest, for it is at these moments when time seems to move fastest heightening the person's awareness that nothing at all is being accomplished.

And so Allan sat there as time ticked forward in its relentless pace bringing with it a determination in Allan to address what was going on. But the harder he tried, the more frustrated he became until he finally called on his guide for assistance.

In these moments of frustration, he knew his energy field was in an uproar and he would probably have trouble talking to the spirit world. He thought this many times in the past when troubles became large in him and he worried about reaching the depths. Yet he always succeeded and so he did this time, not realizing the special access to the depths that was born in him.

As usual, his guide walked him through the confusion he felt, trying to get him to look it in the face, for it is at these times when a person can get so lost in the confusion that they cannot see the trouble that has caused it. He was reminded that the base fear was being seen as some messiah that had come to Earth to save people. He did not want to see this, but this view encapsulated his fear and allowed him to address what it was on the surface.

Yes, he admitted, this was the fear - to be seen as some spiritual leader for he felt quite the opposite. Yes, he did seem to have a talent for understanding his dreams and talking to the spirit world, which was quite all right as long as it remained in private, but the other view of bringing messages to people was not a private endeavor.

His guide pointed out that his messages were written and so could be dispersed in written form and this appeased the reluctant messenger to some extent, yet while he could fulfill his mission in written form, sooner or later he would have to confront a public life that was not to his liking. And so he was appeased in some way, yet the turmoil waited in the wings to be dealt with in the course of time that was allocated to him.

As Allan now confronted his basic fear, he moved more into the flow of the day, and in the evening he was willing to flow with time in spite of his eagerness to lessen the burden he held. He moved to a chair outside his cabin and waited for the night sky. As twilight came and went, his mind shifted from his Earthly concerns to the stars remembering his dream that informed him about his sourcing.

It was a clear dream in spite of having been dreamt more than ten years ago - a dream that still vibrated with a great mystery. He remembered peeling back the skin of the fingers exposing the bone. Bone - the seed of his being. And he remembered how he placed a bit of his finger into the cauldron and it sank to the bottom. It was at this point that the instructor in the dream called all to see and proclaimed that his sourcing was the same as Atlantis.

It had been many years ago when he discovered that the

sourcing which was spoken of was that of the Dog Star. And as the stars exploded in the heavens, he saw the Dog Star rise. The light from the star shone down upon him and he wondered how such happenings could be.

Yes, he had talked to people about it, mostly those who were involved in the latest new age fads, and he felt okay saying this for what they were involved in seemed even more weird than what he was saying. Yet now, in the silence of the night, he looked at his sourcing and wondered about this reality that came from within him.

He believed the truth of his dreams for it was the only time he had absolutely no control over what was given, yet there was always the doubt that the meaning he elicited from the dreams was not what was being expressed, for it was all in the riddle of symbols. But even when he looked at Atlantis as symbolic of a sunken land of old, he still tapped into his own talent, if that was what it was, of having access to the depths of himself.

And he had studied the Dog Star - a system composed of a sun like the one in this solar system and a white dwarf, a dense star heavy with its journey toward what some thought to be a black hole. The tension of these opposites could only source a place where turmoil reigned.

How could he resolve this? What was its purpose? His mind tried to come to grips with these great cosmic movements that seemed to be taking its toll on his sanity.

He closed his eyes and felt the bottom of the cauldron - a place he knew was beyond his own personal depths, far deeper than a personal unconscious and he saw the source of the messages - a place deep in the universe that vibrated with the movement of evolution. And he cried.

7

Today we will discuss something that has long been forgotten. It lies deep within the psyche of humans. It has been successfully buried from consciousness but still affects the world in a negative sense. It is the power of humans' deep seated cosmic sense that we speak. It is not valued because the days when the cosmic connection was honored has long since passed.

People see the cosmic realm as something "out there" in physical reality - a place of space filled with stars and cosmic material that is something to wonder about and try to figure out. They have long forgotten that the cosmic connection which has the power to enliven their petty lives lies within them.

They live their lives independent of this connection and, of course, it is petty and shallow for it is lived out separately from the substance from which they are created. Is it any wonder that they go through life wondering why they are here and must conclude that it is simply to live in a fairly stable environment, raise their families and retire to something other than the work that earned them money to support their lifestyle.

Yet, it seems shallow and it is. For in the depths is their

true nature as cosmic travelers - people who participate in the cosmic movement.

And even as we say that we can see the recipient of our message raise his or her head and look upon the sky above searching for what we are talking about. It has become second nature to look above for humans made a choice a long time ago in their history that what was valued is above.

The head is above; the sky is above. And so they look to the sky for their answers. Yet, as in most things mysterious and filled with potent energy, the treasure lies in a completely different direction than most think.

They believe that just the opposite exists in the down direction for they have long held that this place is the place of the devil. And rightly so, for all their evil and unwanted parts are pushed down in hopes that they will be forgotten forever and so all the parts of themselves that they do not wish to see are present in their depths. They are not deep but at a surface level so they shoot out at times that are least expected.

And so, on top of their cosmic connection is a deep layer of unwanted garbage that must be processed before they can reach the connection to their true nature. Is it any wonder that the ego - the conscious part of themselves - is reluctant to go into the depths for it will find all those parts of themselves that are threats to the ego's sense of control, for the ego does not want the interference of the cosmic realm. It wishes to live out its meager existence in a way that is directed by itself, leading to a life that is shallow but predictable, uninteresting but safe.

Yet, is it predictable and safe? Or is this a delusion perpetuated by the ego to keep it from facing what it really fears - who it is in relation to the rest of the person and who the person is in relation to the cosmic realm? Is it not the case that the ego fears this broadening view of life for it is beyond the mundane existence that the egos of the world have created?

And all around the ego's hold of the reality that it wishes to preserve are mysterious happenings that intrigue the human spirit: visitors from space and miraculous worlds where excitement still exists in the depth of the dark wanderings of the soul.

So, humans have begun to search for what they have lost, yet they look to the sky for answers and get excited by stories of where they might have come from, seeking to find an external reason for their existence. In this way, they delude themselves again not realizing that what they think they see are apparitions that come from their depths as images painted on their eyes in the dark for it is the only way they can see into the depths of themselves.

So, we tell you that what you see is true as far as you can see, for the depths will not be ignored. And if your consciousness is afraid to face the depths of yourselves, they will arise anyway in ways that scare you for they come without understanding, and they come from a place that is upset at being ignored. So, the images threaten at times or come in the night to shock you into realization.

You all ignore the place within in spite of this, for you realize in the night time that much refuse has been heaped over your cosmic connection and the work to uncover it will be painful. Yet, those who experience the work find that the fear is greater in anticipation than it really is for there is great hope in discovering who you are. It is the shallowness of your life that is worse for it is lived in an empty way feeling no connection to the Creator who resides in you.

Your present state brings more pain than you will

experience if you enter your own depths. See your life for what it is and look within for answers. Begin to see the discarded parts of yourself that do not fit into your ego's world and the world of society that all of you created. Find your lost talents that seemed so abnormal when you discarded them. Find your heart and where your passion lies.

Take off the shackles that keep you imprisoned in a reality that knows only what you think you see. It is not as difficult as you imagine but you must make space in your life for the plunge into the depths. You cannot fit it into your already crowded schedule of activities that keep you busy so you will not be tempted to look within. You must reevaluate life and be courageous to try a new path - one that can lead to the miracles you seek.

Yet, you only have our word in this. So, look into your heart and see what it says about our invitation. Be with your feelings and see if tears fill your eyes as you contemplate where your true spirit lies. Believe your heart and follow where it leads.

8

Allan had touched a very deep place in the universe and this was to affect his growth and acceptance of his mission.

Days passed as he struggled and wondered what was happening to him that went so beyond the reality of the world in which he lived. But as the days went by, his astonishment faded to willingness, a willingness to follow what was being given for his tie to the inner realm was stronger than his loyalty to the society that held sway over the general populace.

After a week of struggle, he was now willing to go back to the Guardians to see what was to be given further even though he was still nervous at his role in the drama that was unfolding. He returned to his couch, a place he avoided during his struggles, and opened his notebook to a clean page. As was his custom, he closed his eyes knowing the Guardians were close, and so they were for he heard their familiar greeting.

And so he began what would become the central message of his work. The Guardians explained that he would be receiving the messages that were given to the Atlanteans many eons before, now updated to the language of the current culture.

With this brief introduction by the Guardians, Allan began to receive the first lesson. He had no idea what they would talk about, but he knew that the inner way he had been taught was vital to the work.

And so it was as the lesson unfolded and the Guardians told about the fears of people as they approached the inner realm. It was a dark place for most because what lay in waiting there was not conscious material but aspects that had been ignored for the whole of their lifetime - dark material in the sense of being unwanted and unknown in the light of day.

9

It has been a long time since the planet Earth has been assisted by the Dog Star. Now we come to reinforce the lessons that were given to the ancient Atlanteans. They, as you do now, had much trouble with these lessons, yet we come again to try and help humans accept who they are and find their place in the cosmic realm.

There is a great tendency in humans to focus on what their eyes can see. It draws humans out of themselves and brings them almost exclusively into living in an outer way.

They enjoy building physical things and it gives them great joy to see what they have created, yet finding their place in the universe seems not to give them any joy. It is almost as if they are fearful of where this may lead.

They like to be in control of their destiny and this is their choice. Only now, humans reap the fruits of their choice. There is a darkness descending upon the human race that will cause much pain.

They have ignored all that has been given to them within their inner depths and have used these depths as the graveyard of those portions of themselves they do not want to see.

The depths of each person are contaminated with unwanted images. This plays havoc with their bodies and many are suffering a great deal because of their ignorance of the messages that are given them. They attempt to mask these messages further with chemicals that kill the messages, forcing them to the background, relegating them to a place far away, trying to ignore what is said.

Humans have now divorced themselves from the inner places and so what they create is not appropriate for nor in harmony with the cosmic movement. They are creating an environment where there is no love and no sense of belonging. So, we come again to bring the lessons to them in hopes that they will listen.

We give this out of love and necessity. Consciousness is a necessary force in evolution. When the universe can see what is happening, there is a great participation in the movement. At times, this participation is necessary to make major leaps in creation. Otherwise, the leaps do not take place because of a lack of momentum. Effort is needed and consciousness helps in the movement.

Humans have a great gift for consciousness. Unfortunately, at times this brings with it a great inflation in the humans themselves. They see their consciousness as their contribution to themselves and to the will they create, not realizing that they are members of the All. It creates independence and an unwillingness to listen to what comes from the Creator.

They see the world as their own without regard to the larger picture. This is unfortunate for it creates a separation between the cosmic realm and the human condition. The movement of the cosmic realm has ceased to have an impact on their lives.

They run away in fear for they fear the loss of control. So they hide the connection deep inside and close the door to this inner realm, the only means of communication that they have available.

They delude themselves as they go about trying to create what they think will be an everlasting tribute to themselves, yet it all is built on a weak foundation. It has no substance for it is independent of the cosmic flow.

Only by listening to the messages that are received can humans begin to move back to the evolution that is their potential. Only through the door of the inner way can humans again connect with the All and find their place in the Great movement.

We are here to provide lessons to help humans reconnect to the messages. There is no threat with this. We come in love to nudge you towards a place where you can find happiness again. You walk like zombies to your jobs thinking there might be fulfillment there, but you sense the loss of your own creativity and solutions elude you. You feel lost and unconnected to anything great. We are trying to show you the way, to help you open the door, and to entice you to have courage to change your lives for we only want you to find the love that is offered.

Do not fear what you do not know. Do not set up roadblocks through beliefs that have been heaped upon you. Allow yourself to open to the messages that are offered.

Opening is a word that can be your guide, for you have been closed for many years. You lock yourself out from that which is freely given. Open the locks you have placed on the inner doorways to your soul and begin to listen. It is your true nature that you will unlock. Open to this and you will see again. Unlock the ways that block you from the All. You will see with new eyes and hear with new purpose.

We give you these lessons in hope that you are now ready

to accept what is yours to have. We call you to your own destiny. Have courage and listen.

10

Time passed quickly and the lesson was received. He knew there would be many more lessons to be received each day and he settled into the task.

He had a relatively quiet time between sessions as he was content with the work for it was accomplished in the privacy of his cabin and was something he could do easily.

In the time between the daily lessons, he concerned himself with the more mundane activities of life, but as the lessons began to take on more power, he saw the impact the lessons would have on people. It was a world they didn't know, or if they knew it they ignored it for fear that in the dark realms were the evils of the world.

He knew many conservative groups felt that dreams were messages sent from the devil, and the material he was receiving would probably be rejected out of hand. Yet, he knew others were searching for a way to understand why they were here and he knew if they were open to the lessons, they would receive help.

The difficult part of the message was that inner work was required and he had doubts that people were willing to change the way they lived in order to make time for this essential work. Yet, he saw what the Guardians put forward as the only way he knew to get back on the path that evolution called for. It was obvious that the human species had long since left the path that communed with the cosmos, seeing only the path of the all-consuming ego as it carried out its own plan for the destiny of humankind - a path that led to the extinction of the species.

Other species had already succumbed to the blight that humans had released on the Earth, and more would follow. And it was obvious that the Earth in its movement would not abide anymore of this same harmful effect that came from a species that had so much good to give if they could only have the courage to face the dark side. But he held no great hope for this to happen, yet was willing to participate in what the Guardians desired of him, for it was familiar work.

His thoughts went forward along these tracks, ignoring his own dilemma of being the messenger of such lessons, not realizing that in his participation, he was also being changed and helped to see the necessity of change that was vital for humankind.

11

There is in humans the desire to remain on a path in spite of the disaster that waits at the end of the road. It is a reluctance to be open to change.

There are many people who fear change because they either cannot see where the new path will lead or they anticipate the many disasters this new path will put before them. They build the actual experience all out of proportion.

They see the path into the depths specifically as one that will lead to many deaths and fear the pain associated with that. Yes, there will be pain for it is difficult to address all the ways you ignore who you are.

There are many ways people deceive themselves and to come to grips with this is certainly difficult. Yet, death is part of the way of humans and other conscious beings. It cannot be avoided. Death brings rebirth and it is this aspect of death that is forgotten in the fear of the unknown for no one at the beginning of the journey can know where the death will lead them.

Even in the traditions of modern religions, there is the hope of resurrection. Yet, people still fear the place of death and run away to the head leading lives that are determined from their own logic, which is faulty. People cannot live isolated from intuition and feelings for it ignores the true nature of humans.

There are many stories throughout mankind's history that talk about this fear humans feel about going into the depths. It is the place of demons. Of course it is, for it deals with the parts of themselves that appear to be demons for these parts exist in the dark and their qualities appear frightening in the dark. Yet, once these parts of themselves can be seen consciously, they can assist the journey of the human and, in the larger view, that of humankind themselves.

But rather than take the chance of seeing who they really are, humans run away from the depths hoping that the demons will not get them. In this action, they insure that the demons in them will surely chase them down.

Only in consciousness can the demons be seen clearly for who they are - parts of themselves that need to be addressed and felt so that their true nature can combine with other parts of themselves. In this way, the demon is experienced and can be seen as an asset to help humans deal with their own fears and doubts about their own nature and spirituality.

When the demons are left in the dark, they exhibit only the dark portions and invade the human's world with only their negative qualities. Consciousness can bring out both the negative and positive. Yet, at first look, the human fears the negative for the human anticipates that nothing can come of this that is good and thus fulfills this fear by allowing the demons to stay in the dark, always pursuing them as a threat.

Only the courageous who go inside to confront the demons have the benefit of seeing how the depths can assist the human ego to relinquish control so that the true nature of the human can be realized. The soul is then free to connect in a conscious way with the celestial rhythm.

Yes, there are many paradoxes here that must be faced. Listen to these words for only by going within can humans attain their true spiritual nature.

12

Days passed and Allan could see the messages take on a power he knew could not be ignored if people would only read what was being said. And this brought back his own struggle of delivering messages that would be looked at with disdain by some, but with openness by others who would see the wisdom of what was being imparted to them.

And so, he again saw himself as a messenger from a foreign land arriving on the shore of this world that was unfamiliar with the ways of the spirit world. How to deal with this was his major concern. Would he be laughed at as some strange character like the hermits of old? Would he be ignored as someone who had obviously been affected by long stays in the woods, out of touch with the reality of this world?

Then he was struck by the opposites of the two worlds he lived in. One world was made up by the egos of humankind, and the other world was inhabited by beings who lived beyond the physical - a world where information was readily available if you could but tap into it, and that he had.

And as strange as it was to the majority of humankind, it was the world to which he had grown accustomed and where he placed his trust. He had no trust in the ego's world, for it dealt with deception and manipulation, all to fulfill the economic goals of growth beyond what was appropriate. For the system required constant growth to stay alive. Consumerism had been like a god who sought more and more allegiance to its own agenda and consumed more and more of people's time to satisfy its insatiable hunger.

He distrusted this world knowing the shallowness of its methods. And so he could see himself as an emissary from a foreign world as he approached this society with these messages, and that somehow elevated his overpowering concern as to how he would be seen.

Yes, he knew society's desires and goals for he was a good student and had learned their lessons, but he also owed his allegiance to a world that went beyond that world view. It wasn't so much that he had one foot in each world, torn between differing agendas. No, that wasn't it at all. He had both feet planted firmly in the world view that allowed for the unseen, but that also included the physical, yet not in the same way it was seen by current society as the only view possible.

It was as if his world view had horizons that extended out further, beyond the view of purely the physical, not in any linear expansion, but in an expansion that was within him. It was an expansion that encompassed both light and dark, the outer and the inner, the mundane and the cosmic, all flowing together in a way that seemed more appropriate than anything he saw in the realm of cities. Here all activity seemed to be focused on the growth of ego power and control over a world that wasn't understood.

His musings brought him to an unusual place, for he had always felt the tension of the two worlds he thought he straddled. He now could feel a strange strength emerging from his gut, a strength in his world view that no longer seemed so full of contradictions, but full of expansion encompassing all that he believed, not ignoring the culture's view, but . . .

He couldn't quite get there - something different that didn't seem to be such a threat to his position in the world. Maybe firmness in his beliefs, a belief that knew the wisdom of what the Guardians spoke. His stumbling around the avenues of his musing had brought him to a new place, a place he couldn't quite recognize, but one that seemed to ease his soul allowing for a crack in his fierce protective covering that sought to preserve some semblance of normalcy.

Yes, a crack in the armor had emerged that had possibilities to resolve what was bothering him so much when he contemplated bringing the message to the public. What it was he didn't know, but there was somehow comfort in his meanderings around the labyrinth of his own thoughts. And so he was content to be quiet for a while and let what was moving in him take on some further consciousness, giving it time to ferment as he puttered around the cabin in the quietness of twilight.

13

In this life that humans lead, there are many ways to see. They can believe that the physicality of the world is the only reality that exists. This creates many dilemmas for their spiritual life.

It requires symbolic associations that become only further imaginary physical images. It requires people to invent spiritual practices that are physical, like congregating to speak prayer and do activities and contemplate spiritual words that have no other dimension than what is said. This is a very limited spiritual way and hinders humankind from moving forward into an evolution that can bring them to greater insight and understanding of this universe in which they find themselves.

So, the view that they see of the world is very dependent upon how they live out their spirituality. If they have a view that is only the physical, then their spirituality is played in this physical world with physical activity. If they see in a broader way, they begin to see all that is available to them to assist their spiritual movement.

This broader vision goes beyond the physical, or the surface of things, and can go into the living energy of what is there, for all living things are but faces of the universal One.

It is the one and true God and it is the All.

The All relates all of life through the forces that are contained in the universe. There is no way that the All could exist if everything that appeared to be physical were all separate living entities.

Think about this. Could a planet exist by itself without having relationships with the star or stars of its system and with other planets? They all dance in a field of relationships, all dependent upon the interactions both within the solar system and beyond as the solar system interacts with the dance of the galaxy.

Humans know about these forces in a limited scientific way and so do not see that the play or dance of the movement is a great spiritual event. As the movement of heavenly bodies contributes to the spirituality of the universe, the dance continues to the farthest reaches of the cosmic realm. Galaxies and other heavenly bodies all move to the music that brings forth growth, life and death in a harmonious crescendo of evolution.

And just as the infinite stretches of the universe move and dance, so does the infinite within. For within all living things, there is the same dance being carried out. It is why the ancients knew that the human contained the universe, just as the universe contained the human.

For within the All, there is this flow and movement in every living thing. So, there can be no separation in the universe. The flow and movement know no boundaries. Movement does not stop at the surface of a thing, for there is no surface. It is an illusion that one sees, and the wise humans go beyond this physical limitation.

14

Allan had moved to a place in his musings about his turmoil where he was beginning to see a different picture. No longer was he outside of society as he had thought. He was within this society but in a place where he could see in a broader way than most. He saw society as it was - concerned not with the whole picture but only the portion that dealt with the physical.

Even the modern religions had little to do with spirituality for they compressed the spiritual messages into everyday living, focusing on the physical concerns of behavior and the fear of physical death. There was little concern about the way of the cosmos other than some playing with mysterious symbols that lost their power as they were converted to physical reality.

Allan's view of the cosmos was a place of living energy where all intermingled in a dance of movement in which he shared. He could not only feel the movement and recognize when he was not a part of it due to his behavior, but he could communicate to the depths where the love of the universe pulsed in a rhythm to which he could tune. He knew this in his heart to be true and his dreams confirmed what he felt.

He believed his dreams wholeheartedly as the truth that

spoke out in a sea of confusion. Each day during his turmoil he was encouraged by his dreams to continue on, each dream addressing a concern of his that was not well founded - his concern that he was not receiving the Guardians' messages correctly, his concern that he was lost and alone, and his concern that he had lost the path.

All these were addressed specifically by his dreams showing him that he was on the spiritual path, that he was surrounded by the people in his inner group who were ready for some spiritual breakthrough and that he was connected quite well to the seat of cosmic information.

He knew this to be true for his dreams had spoken it and so he tried to relax and let the movement that would not stop continue unimpeded by his own troubling concerns.

15

Many do not understand why we are here to encourage the people of Earth to plunge into the depths. These are the people who refuse to see the larger picture of life. We say to these people to look around at the world they are living in and see it for what it is.

Is there meaning in all that you do? Do you see the purpose of your life? Why is it when you go to work or to play, you are always searching for something that fulfills your life be it money, prestige or a sense of self-importance?

Can you agree that there is an emptiness you feel or are you still thinking that if you continue to search within the confines of the way of your world that you will find who you are? These are important questions to contemplate. If you are fearful of the answers, remember that it is your feeling and the feeling is telling you something.

Listen. Be with the feeling and allow the messages associated with the feeling to be revealed. This will help you get in touch with what you are doing to yourself.

It is all choice. You can try to blame the way of the world that has been created for you, but it is you who have made the choice to follow the way you have been taught. You cannot blame what you do on others. It is your choice that determines how you behave.

Yes, these are difficult words but the consequences of following a way that does not bring fulfillment is even more difficult for it locks you into a way that ignores your very nature as a spiritual being.

Are you surprised at this statement? Yes, it is surprising to one who relies on the material world to provide all the fulfillment that is desired. This is quite a load for the material world to carry.

Why is this? It is because you see things as having no spirit so you view the material world as a place where ownership of things is the important goal of your life whether the things are possessions or money. It is this that you think can fulfill you, yet it has no spiritual aspect.

Can a person who has a nature in the spiritual be satisfied with the rewards of the material? Yes, you say, but you go to your church every holy day and pray to the one God for guidance on your path. Yes, we know. But the place where you seek answers is out there in the universe and the answers are closer than you think.

If you are a spiritual being, you contain the spiritual. Where does this spirit lie? It is within you. That is where your nature is. And your nature derives itself from the cosmic realm. These connections are also within you and so are all the emotions that you feel that can activate this spiritual nature within you.

We tell you that you are going through life as if you are asleep. Knowing the external material world is not being awake for we speak of being awake to your spiritual nature. To be awake is to listen to what is constantly rising to the surface of your consciousness. Consciousness is the process of awakening to your true nature that lies within you.

So, this is why we are here to encourage you to take the plunge inside you. All the answers are there. You are now focused externally. This is the wrong direction initially. How can you see the messages that are constantly being presented to you in both the outer and inner world if you have no experience with the source of these messages?

The source is within you. Look here first. Then once you begin to understand the messages of your soul, you can then see the outer world with new eyes.

We tell you this to try and help you get into who you are. There is no mysterious overall plan that all humans are to follow. The only mystery is within each person. It is to this realm that we point you.

Begin to feel and see where that leads. Learn about your dreams and seek help in understanding them. Be aware of the messages that are constantly bombarding you - not the external messages that seek to rob you of your god-given time to learn who you are, but the inner messages that call to you for attention.

Yes, the path of these communications is rusty because of lack of use. Yes, it will take effort to begin to listen and to open the doors you have closed to your own inner realm. But think about your life.

It is to be a fulfilling life, yet do you feel fulfilled? You are placed here to be nourished by your true spiritual nature, yet do you feel nourished?

We ask you to seriously contemplate where you are in your life. Make the time. Look at your life and find that first sense of your spiritual nature. It is freely given. All you need do is open the door to your inner realm and allow the messages to come. We encourage you to do this for your own sake and the sake of all humankind.

16

Allan passed the days in pleasant undertakings that he enjoyed - doing some necessary chores about the cabin and putting things in order for he felt deep inside himself that a move was in store for him. And so he busied himself with these activities as he continued to receive the messages from the Guardians.

He could see how the messages got to the heart of the problem in people who ignored their depths. They spoke of fear, a fear that had been ingrained in them over many generations - a fear of reverting back to a savage state with an impetus to stay above the depths for fear the savage part of themselves would come alive.

He knew the fear, yet he had entered the depths and found his instincts to be helpful. Rather than a detriment to his spiritual journey, they were an asset to his understanding about the way of the universe, so he knew the fear to be unfounded.

This society favored the ways of the head - the thinking realm of the ego that believed in logic and reason, not wanting any help from instincts or intuition. Yet, many examples from their great thinkers always referred to the intuition that came to them out of the blue. Such intuitive conclusions took great effort to prove with logic but the insights came with ease. He wondered why these lessons were ignored for they came from the heros of this age.

Perhaps the fear was too great for the majority of people to plunge into the depths and try it. They feared monsters, he presumed, that would devour them, monsters that lived in the dark place. He knew there were monsters but only of each person's own making, for in ignoring their own inner selves, they relegated the dark portions of themselves to a place where the monsters screamed to be noticed and acted menacing for they were parts of themselves that were unknown.

This was a strange world where people walked around in fear of themselves, preferring to focus on external matters as if the material goods of this world would bring them to a place where they could find satisfaction and contentment. Even their religions promised happiness in a place that existed in death which was physically located in the heavens. Such was the way of things.

As he mused about the state of spirituality in the world, Allan thought of the wisdom of the natives of this land and how they could see the spirits and the interconnectedness of all. Yet, even they were torn from their beliefs for they were labeled savages when they voiced this view of the world. And so they turned to the white man's ways to try to find some place to live that was viewed worthy by others.

These meanderings of the mind brought Allan to many revelations. Here he was holding this broader spiritual view, willing to act as a messenger for the spirit world, and so concerned with how he would be seen. He, unbeknownst to himself, was falling into the same trap that convinced the native peoples to forego their beliefs and join a culture that was lost in the world, yet seemed to exhibit a power that engulfed the planet.

The power that people admired he knew would not hold a candle to the power of nature. Yet, even with recent demonstrations of this power to destroy the ego's way of life in an instant, people still cowered for fear that by leaving the ego's ways they would be considered insignificant or below the intelligence that the ego proclaimed.

He could see the shallowness of his own thinking, getting caught up in the same concerns about looking stupid, or savage, or lacking proper reason. He knew that he had thought many moons about what he believed and what he had been given. It was not trivial and was, at times, far beyond his capacity to understand.

He remembered the turmoil as he tried to learn the language of dreams, and tried to understand the way of the universe in its interconnected dance. He also remembered times when he worked in the culture as an analyst, building computer databases using techniques that interrelated activities into meta structures. The interconnected diagram showed how a whole organization could be understood with one picture.

The structure of data interconnecting with other data came from within him as he talked to people in an organization about what they worked on. And he could see how all the different tasks they performed could be understood as a whole pattern of information. But those who favored logic and programs were more content with separate processes, separate data files, fragmented views of what was done.

He knew the shallowness of their views, for he had extrapolated those years of building information structures into a picture of the pattern that existed in the universe. He saw this not in any structured, mechanistic way but as energy patterns moving in similar but more intricate relationships flowing in what some considered to be the empty space of the universe.

Yes, he had struggled many years to try to make sense of this, and now contained a semblance of knowledge about how such patterns could move and affect the smallest, most insignificant action. He had a sense about how the universe in its movement of evolution could swirl the gaseous materials of stars in the same way it could move a person's heart.

He cursed, in comparison, the linear ways that society was built on. He was trained as a systems analyst able to manipulate mathematical models that he knew could not even get close to what was happening as the movement of the universe danced in ways that brought all that was interconnected each to a harmonious place.

Using the way of logical human thinking to describe the world, people chose to look at this or that activity, but with their methods they could never handle all the interconnections that existed. And so they deceived themselves but told a great story about how they were on top of what was happening.

He knew that the understanding of the whole had to go beyond the logic of these methods. For the larger answers resided within, within the far reaches of the depths inside where understanding could be achieved not as mathematical equations but in the movement of the heart. It was where the information could be understood in a place where spirituality merged with the physical, and where thoughts merged with insights. It was here in the space of an instant that understanding could be achieved if one was just willing to listen and put aside the power of the ego to control.

And he was worried about how he would be seen? Yes,

certainly not as a scientist, nor as a religious fanatic. But also not as a lunatic, for he had done much work understanding the inner way and it was not to be taken lightly.

Yes, he thought, he could point to this understanding that flowed out of his work with the depths and yes, he could see the pattern of a dream in all its glory which befuddled the most logical mind. But this wasn't the purpose of the messages to show his prowess in the face of the mighty cosmos. No. It was to lead others on the same path to the inner realm that he had taken, to come to their own conclusions about themselves and their place in the universe. He had only the answers for himself for that is how the universe accommodated all living things - each a part of the whole, but each unique in their own contribution to the movement of evolution.

Allan stopped his musing wondering where he had just been and why his thoughts had taken this turn. But he now saw himself not as the outcast, the strange one who needed to apologize for his way, but as a leader pointing the way into a new world far beyond the logic of men. And as he felt this new confidence build within him, he wondered if he wasn't going too far in his own power, inflating himself beyond what was there. He knew he had to rest with this and let it all resolve further, but movement was occurring, even as the chores of the day carried him forward in time, allowing the space to deal with his own mission.

17

We wish to concentrate today's lesson on the strength of the ego to hold onto beliefs that hinder a person from entering a place where they can see the world with a clear eye.

The ego of the world - that is, the collective ego's of humankind - has created a society that depends upon certain behavior from her citizens. One strongly held need is the ability to see the value of the society's view of the world. This is based on clear, rational thinking.

There is a need expressed in this society to explain what one is doing and what one wishes to create according to the way this society sees the world. There is, in other words, an accepted language that one is supposed to use. It is built upon logical foundations, cause and effect.

This language is based on logical extensions of what is already in place. If one wishes to add to the growth of society, one needs to explain the growth in terms of past accomplishments. This is the acceptable practice. If a person sees a need, it must be expressed as a continuation of what has preceded this new need. In this way, growth is seen as a process of building upon a solid foundation that has already been laid.

This progression of accomplishments establishes a view

of this world that people live in. It is a view that is aware of what humans have built before and how the new ideas add to this to increase the growth of what has already been started. Thus, growth is defined as the expansion of what has gone on before.

This is one view of the world, and one the ego prefers. This says that what the ego has created is everlasting.

Now, there is another view of this world that is also everlasting that is held by the inner way. This is not based on a solid foundation but on the ever-changing pattern of possibilities. Rather than a linear expansion of what is present, the inner realm sees a world of ever possible movements in a pattern that is affected by all that exists. Thus, it is not confined to the world view of that which has been created by humans.

It includes what humans have created, of course, but also includes all other natural patterns of development that exist in the universe. And in the movement called growth, all patterns work together to move the All forward.

This is quite different from the ego's world view. The ego's view is based on what the ego has created and the continuation of this until more powerful egos overthrow these ideas and begin to perpetuate the new ideas of what is important to create.

Thus, it is important for the ego group in power to create stabilizing influences to maintain their ideas of what constitutes forward growth. These stabilizing influences are based on always building on a stable foundation, meaning the foundation of beliefs that are held by the ego group in power.

Now, what are the beliefs of the cosmic realm? Here there is a basic belief in the natural flow of living energy to create an environment where all contribute to the whole movement.

So, in order for living entities to exist in this flow and contribute to it, they must be aware of the pattern of movement that is currently happening.

At times, this movement creates great surprises and makes great leaps forward - forward in a sense of growth that contributes to some sense of spiritual enlivening. These leaps can excite further growth in ways that have not been perceived.

Such was the creation of the life form called humans. This was a great leap for it held out the possibility of consciousness - a consciousness of the patterns of possibilities. In other words, it became possible for the All to know where the growth was heading rather than to simply exist within the growth.

It was like the possibility of an eye opening, seeing itself for the first time in the reflection of what was going on. For consciousness is like this.

As an example, if a human is in a process of spiritual growth and does not know it, the growth still happens but the human cannot see the meaning in it. The growth continues but has no purpose that can be seen.

Then the human sees with an open eye the growth that is attempting to happen within their inner self, and the growth now takes on a more vigorous excitement. The purpose of reaching some sense of themselves takes on importance. This heightens the experience and consummates the growth in a heightened sense of awareness. Consciousness feeds the growth, and the growth feeds consciousness so that a new level of self-awareness is attained and more of the purpose is known. This is the best we can do to explain this in current human terms.

And this is the view of a universe beyond the ego. It is one in which the ego has a place but is not in control for nothing is in control but the pattern. This is threatening to the ego and the basis it has built for its own world view: a solid foundation that encourages linear growth.

Patterns are not linear nor do they have a solid foundation other than the basis of change and movement. This is terrifying for the ego to contemplate such a world, yet it is the natural world that exists. It is because of this inner realization by humans that the natural world is so different from the controlled environment they attempt to create, that the ego tries to stay as independent as possible from this cosmic flow.

Yet, we state again that it is what exists. It is the environment you are born into and it is now time to address it. And you are born with a capacity to understand the pattern: dreaming.

Dreams are not there without a purpose. Dreams allow the next evolution for they can bring you consciousness rather than the logical reasoning you have so long cultivated. They bring a true awareness of the pattern of movement that surrounds you.

So, we say, it is your destiny to accept this inner path if you desire to continue on with the evolution of the universe. It is your time to participate. Do not let us down. Think about these words and see your destiny.

18

The movements of his previous musings were startling to Allan and he needed time to understand better the path he had taken through his confusion. As he thought about how a person, who saw in a broader way, walked through this society and how his vision affected his walk, he began to see what really was going on in himself.

He knew in the past he had put his inner talents in service to the work of society and he saw how frustrated he had become as he always seemed to come to a brick wall. He now understood that the wall symbolized the limitations of looking at the world only in a physical way. He saw that when he applied his broadened vision, as then exemplified by his intuitive skills, to a world that only valued the physical, he was frustrated. The inner vision saw beyond this limitation, but he was only afforded opportunities to manifest within the confines of the culture's world view.

He learned a great lesson in this. He saw the inappropriateness of putting the inner talents in service to a world that could not see in a broader way. This limited his vision, encapsulating it in a smaller view. People simply wanted the benefits of his sight without broadening their own view. They took the pieces they needed, and discarded the interconnections that held the broader view together. These were the fragmented ways he had seen and detested.

And so, in his musings, he learned that the ego's world could not be served fully by the inner way because the ego always looked at the world in fragments for it could not comprehend the broader vision. It was a necessary step in his understanding, for he realized the change had to come from the ego as it loosened up to learn this broader vision of the world.

He also realized how difficult this was, for it meant that each person in society would have to address their own limited vision as it was lived out in daily life. Each ego would have to admit that it saw in a small way as compared to the vastness of the inner depths and this destroyed the arrogance and shallowness of that part of them they relied on, and it was painful.

19

We are ready to get into more details about the inner way. Today we wish to present the path to the center place.

The center place within a person is the place where the vertical and horizontal come together. The vertical is the cosmic connection, connecting the Sky realm with the Earth realm. The horizontal is the plane where everyday life is enacted out.

So, in the center is where the influence of the cosmic realm is felt directly on your everyday life. Remember. This is an inner place where the two energies of life come together. It is not something you will find in the outside world.

To reach this place where you can be in communication with the cosmic realm and live harmoniously with the evolutionary movement, much work must be done internally to bring all aspects of yourself to consciousness. Some have described this as peeling away the infinite skins of the onion. This is an appropriate symbol, for the onion is a bitter vegetable that can bring tears to your eyes.

Each day when you work with your dreams, you will discover which aspects of yourself you need to be aware. This is a difficult task, for your ego needs to learn that it is not alone.

The ego in acquiring its strength becomes almost like a false god thinking that it can move mountains and attain great heights. This is a necessary part of acquiring a sense of self. This is not the inner Self but more associated with the physicality of your personhood.

You learn to distinguish between you and others. This is difficult as a child and much effort is expended to differentiate you from the energy of the universe. This is the beginning of consciousness. It is a way to see uniquely, but at the same time you forfeit your connection to the All. This is how it is. It is neither good nor bad, just is.

You then begin to learn all the things that help you become a contributing member of your culture. Much time is spent here. You are taught skills that you will need, appropriate behavior, history of your culture's way of seeing the world and acquire the necessary motivation to fit into what society has to offer.

Since your current society is focused solely on the external, this training has a deficiency for it drives you further away from the cosmic realm and reinforces your childlike view that you are separate from everything else.

So you join society as an individual who is expected to grow in stature and wealth acquiring the materials needed to fulfill a life of pleasure and reward. This is the way of it. Much effort is expended to do it and as you try to acquire a sense of worth, you move further away from your center place.

When you discover that this path is not fulfilling and you realize you have come a long way without getting any sense of who you are, there is much work ahead of you. You are then called to begin the journey back to the inner center, now with a strong consciousness of the world in which you live.

You are now ready to encounter your depths.

This is not to say that all the training and the excursions into what your culture has to offer were necessary. However, you had little choice in this, for this is the way your society offers to a new human. So, you have more work to do, for many of your aspects have been buried in your depths in consolation to what the world wants and now all these aspects of you need to surface.

Your goals and aspirations now need to be tempered with the realization that your great mission in life lies somewhere within you and what you have long sought for may not be appropriate to who you are and what you are here to do. It is quite a difficult doorway to go through to know that you are a member of a universe, a galaxy, a planet that is flowing in a great surge of evolution, and you, as a unique individual, are still connected in a very deep way to this flow, in spite of your earlier training that you are an independent being.

Yes, it is quite a shock to realize that the All is all related and moves together while preserving the unique expression of each entity. For each living being is one face of the All, and all contribute to the movement.

Your greatest gift to this movement is your consciousness, for if you do decide to go within and discover who you are, you bring consciousness to the universe in your journey. This is one goal in the All - to know what it is. And you have a definite mission as a collective species to bring this to being.

How much more fulfilling is a journey where you, the planet, the galaxy and, yes, the All can all consciously know what that movement is. And so, all of you have a great mission to bring consciousness to the world as well as to act out your true nature as a unique participating human in the evolution of the All.

It is at first a difficult entry to begin to learn your other aspects and meet them in your dreams, and to meet those inner people who do not contribute to your well being. Yet, if you will see the bigger picture of what this is all about, you will not have a tendency to think you are alone in the journey. All the unique faces of the universe look on you and support your efforts.

You are surely not alone. For in the whole of the cosmos, there is no one more important, for all contribute to the whole. Can the All favor one over another when all are essential? Realize your importance to the All and begin to see that your life is much more than you imagine.

You are a spiritual being who is essential to the universe. Can you believe that seeking fulfillment in other places will match the destiny which has been seeded within you? Open the door and seek this center place within you.

It will not come easily, nor quickly. Yet, within your lifetime is the promise of great consciousness. Begin to awaken yourself to something that is offered freely - your true nature.

20

Allan continued on the next day trying to understand how all these changes were possible, and he knew the only avenue was to work with dreams, for this work involved broadening the ego's view to accept the other parts of the person of which the ego was unaware. And in the personal broadening of vision there was the beginning of the opening - an opening that would allow the person to see beyond the confines of the view he or she had been taught.

And so he was brought back to the inner way as the path needed for further continuance of a species that had come to the end of its vision, for in its fragmented view of the world, it failed to see the interconnectedness of the All. While he had not an easy picture of what was happening, at least he knew that he was seeing a world differently from most.

He could understand his own frustrations and he realized his anger at society was an anger at the situation that put his view in direct comparison with the view the majority held. And his frustration at not being able to manifest what he saw was the result of trying to apply it to a world that had this narrow view.

This bothered him for he saw the work that had to be done by each person. He realized this was not a change that could be implemented by instituting a different society, for no matter what was done politically, it could only change the methods within the same physical world view and this would accomplish nothing. The limitations were there not from the rules and regulations that governed the way people were encouraged to live, but from the very foundation of how each person saw the world they lived in.

It was more basic than any way of governing, for it delved into the very ideas of what the world was. It was prior to government, and was contained in the heart of their spiritual beliefs about what the universe was - a physical place that begged for sanity in the logic and reason which formed the basis of its beliefs.

And in these beliefs was the rejection of the unseen for this did not conform to the logic of its view. It was the foundation that society was based on, a way of thinking that confronted a broader view without any means of merging it into the ideas society held dear. For the broader view had to deal with an unseen world that went beyond the logic and reasoned ways which kept its believers trapped in a smaller view that was inappropriate for this time.

Allan became depressed at his own conclusions, for he did not believe that this society was willing to do the inner work to bring them to a place where they opened enough to accept a realm that went beyond the physical. It was a terrible realization for him and he walked aimlessly again into the woods wondering why he had such an impossible mission to carry.

As he walked, he recalled how he had lived in the world and used his inner eye with a lack of consciousness, for he was unaware at the time of what faculties he was using. He did recognize this now. Then, if anyone had asked how he could see the patterns of things as they came together within him, he could not answer, or when he did attempt an answer, he could only say he got the information from the people he questioned.

He knew other analysts wanted a step by step method to recreate his own interconnected diagrams, or his understanding of factors beyond the mathematical models. He also knew his lack of logical method frustrated them, especially when they could see the value of his conclusions.

Yes, his conclusions all had a semblance of logic since he wrote this way to convince others, yet the method was not clear and it frustrated those who had to work with him. Perhaps, he thought, if he had been able to explain it, it would have been better - better in a sense of being accepted. But now he knew his explanation would be about methods that went beyond the linear logic that was prized. He realized he would be in a worse place with those who sought his methods for it would have only ended up admitting a world view that used the inner realm as the source of what he saw.

But now he could talk about the inner realm with its language and ways for he had spent much time contemplating it. On the other hand, how could he talk about spiritual things in places where people prided themselves on their scientific or logical methods that sought objectivity rather than the inner spiritual places of the heart.

No. Consciousness about what means he was using would not have satisfied his detractors. If anything, he would have given them the ammunition they needed to show why he wasn't to be trusted in spite of the accuracy of his conclusions. And so he mused about his past experience in the world of business and further complicated his own depressive thoughts about what he had been given to communicate.

Allan still had not come to grips with his own problems of delivering the message to a world that had no room for it. He still saw his mission as daunting only in the external enormity of what had to be accomplished, for he still saw the task as one that would have to change society and allow for a larger view. Once he stayed with this statement, he would see that this was not the task. He would get to it soon and so the spirits waited as he walked in confusion, seeing himself still as a messenger to society.

21

It has been a long time since we have talked to humans so we are not exactly sure what it is that they need to hear. So, we are telling them the truth about their nature and it will have to do.

We would like them to take this to heart, yet the strength of their commitment to external things is very strong. We can only share the lessons that were given at the beginnings and hope that this will excite something within them to want to find their true Self.

For so long, humans have been separated from their very center. They are moving in unrelated ways, doing, doing and more doing. Yet, in all their doing, there is no connection to their source.

They are like flies buzzing around a piece of meat, not sure of their surroundings, but certain that what they are doing will eventually sustain them. Yet the meat is tainted and it will surely kill them spiritually.

It is of major consequence that this does not happen. Many are dependent upon the continuance of this species. Many are dependent upon the great influx of consciousness that can come from this part of their galaxy.

It is vital that movement continue in the evolutionary

process of the cosmos. It is on the verge of collapsing and only this great influx of consciousness can reverse any tendency to fall back on itself.

And consciousness must serve the depths. The depths are where movement is consummated. The movement begins here but it needs the awakening into the light to keep the movement going. Humans contain access to these depths and can bring to light the evolutionary movement and succeed where others have failed.

Do not think that humans are without place in the great scheme of things. All can be achieved with their participation and much can be lost if they choose to continue on in their separate way, living as if they do not matter to the All.

This is not the case. They are vitally important and we hope they take this to heart. Many species on Earth have come and gone and none have more potential than humans to be participants in the great drama of evolution.

And it is this time that is ripe for their awakening. We only hope that they can hear these messages and pay attention for the good of all they hold dear in their hearts.

We have added this digression to the lessons as a reminder to all humans that these lessons are not given lightly nor are they delivered as entertainment or for amusement as they go about their everyday lives. It is important that they look at their shallow lives and realize what they are doing. All depends on their reevaluation of how they live. Nothing can be accomplished if they continue on this path of self-destruction.

They have momentum on this path and much effort is needed to stop the rush toward destruction. It is like trying to stop a large boulder as it plummets from the top of a mountain to the valley below. Realize it can be done but it takes personal effort.

Each person needs to address the direction of their life and begin the change to focus within, for it is here that the answers lie with the power to reverse the course from destruction to life. Be not afraid for you have the courage and the ability to do this.

It is not easy but you have the abilities to follow the inner way and change the path. Put your effort here. The rewards are much greater than you can imagine. You will finally realize the potential that resides within you.

Stop chasing the hope expected in external enticements. You know from your past efforts, this all fades once you reach the external goal you have for yourself. You disappoint yourself each time, but you still seek the external gold.

The gold is within you and can be found by digging into your own depths and uncovering who you really are. Make this your goal and you will not be disappointed again.

Look around at your external entrapments and see them for what they are - vain attempts at glory that have no lasting effects and that pale in the full light of consciousness. Delude yourselves no longer and listen to what we say.

22

For the next few days, Allan continued to receive the lessons from the Guardians while he thought about his relationship with the society into which he was born. He knew he had a compulsive desire to achieve and knew that his work had been respected although he was constantly involved in controversy. What he recommended always seemed to be out of the mainstream, yet he found arguments to sell the ideas to those who were in charge.

He saw this in himself and at some level he was pleased with what he had done, but he also saw that what he personally wanted to achieve in society was not to be. For each time his views were noticed and they wished to reward him for his good ideas, he found himself promoted to places that ceased to be a creative place for him. It was a place of budgets, and consuming goals to keep business going in spite of the ways that were used.

This did not sit well with him for he detested this separation from the creative tasks and abhorred this all consuming desire to make money independent of other goals like helping the client and finding the best way to achieve the outcome they desired. At the higher levels, these motivations were not what was important. Instead it was preserving the business in a way that fit in with the goals of expansion and increasing the holdings of the corporation, or institution.

He could see the need to do this on some level, but at the same time he fought it. He knew these were acceptable ways but as he was molded to adopt these ways, an inner part of himself could not abide them.

Yet, he desperately sought the approval of whatever management existed in the particular job he performed and it didn't matter whether the management was that of his own place of work or the management of the people he served as an analyst. In fact, he tended to favor the client which made him popular with those he served, and this sent messages back to his own management that said he was a good money maker, for the client wished to continue on with his services.

And so he pleased both his own management for fulfilling business goals, and the client's management for providing solutions. Is it any wonder that he was singled out for promotion? And he felt honored by this.

But it wasn't long before he felt the falseness of what he was doing. For as he advanced to management levels, he had to become the defender of the business while the services provided were moved to a secondary place for consideration.

This he didn't like. He was there to be of service to people, not the keeper of the business. And so he rejected their constant demands to meet business goals, to motivate his people toward business ends, and to keep clients happy while making money at their expense.

He had been trained for all of this, yet he excelled as the servant of the client, and became angry at the tasks of management. And so he jumped ship in hopes that the next job would be different, but the cycle continued. He excelled again in serving the client, was noticed by management as someone who would advance the business and was groomed for management. And when he was promoted, he was honored.

He could see that he sought the recognition and rewards of management, but when he became a part of their group an inner explosion was set off that he felt deep in his heart. He could not abide their methods, but at the time he had no conscious sense of why. It was just an unconscious urge to run away as fast as he could.

23

We will continue with ways that contribute to the confusion of people as to their own true nature. Apropos, we will talk about the pull of the collective society on an individual.

There is a tendency of people to want to go along with the group. The saying, "there is safety in numbers," seems to fit this society. If a particular occupation seems to catch the fancy of everyone, then people will flock to this occupation whether they have an inner passion for this work or not.

They wish to cash in on the popularity of the job so that they will look good in other people's eyes. If people admire and seek after the job, then they will admire and seek after the person in the job.

Of course, all this goes on in the unconscious of a person and the popular occupation becomes a symbol for the self-worth they are seeking. Although, rather than consciously seeing the symbolic significance of seeking a popular job, they seek to make the symbol a reality in their life thinking this will bring them self-esteem.

Many are sidetracked in the pursuit of their true nature by choosing an occupation that is not appropriate to who they are. They learn the ways of the occupation and heap layers of behavior upon themselves hiding their true nature.

This is unfortunate for it only creates a great gulf between who they wish to be and who they truly are. So, again they focus on the external image of the job rather than the inner image of themselves.

This is not just the behavior of a few. If the job is not available to the person but he or she would still desire to be seen like this in order to garnish self-esteem from others, he or she will put on attributes of this occupation as a facade covering who they truly are. They will tell stories to invent this demeanor as appropriate to their position in life.

That is, if they are not, for example, a lawyer, they will invent stories about similar jobs that they are associated with or attempt to form bonds with lawyers so that they can "rub elbows" with them. This is far worse than the one who actually becomes a lawyer, because it begins to add to the delusion of their life and moves them into a fantasy world that has no direct connection with reality.

There are people who can live this delusion for their whole life time and never discover who they are. This is a disaster to their own inner life, forcing it further and further away from themselves, creating many negative emotions that can explode outside wrecking havoc on their own behavior and on those associated with them.

This is not to lessen the destruction within the one who succeeds in becoming the job that is popular. While here the delusion is not as great, it still creates a path that may be adverse to the necessary journey of the person to his or her true nature. Many people are affected in this way, some more harmfully than others, depending upon which occupation is popular at the time.

We tell you this to impress upon you that your job or occupation in life needs to be a reflection of who you are. It is a manifestation of your true nature.

While all in the universe is interrelated and dependent upon each other, an individual needs to express his or her true nature to be expressive of the diversity that is part of the natural way. Each shows a unique face that contributes to the whole.

It is only those who have discovered who they really are that portray this true picture. It is the nature of the universe to show all its faces so that all will know who we are. For it is in seeing all the faces that we can define better our own uniqueness.

See this for yourself. When you are in a group of different people, do you see your own differences? Yet, when you are in a group that has not found their own uniqueness, are you equally lost?

It is only when you have found your unique self to which you are called, that you can see the uniqueness in others. Then you are not stealing their power or nature but are content within yourself and thus see others with a clear eye.

See if this isn't true for you as you mingle in a group. Are you confused about who you are, floating from one facade that you see to another thinking that this may be you?

Realize that the only way to truly discover who you are is by going within yourself and doing the inner work. If you do not, you will join the majority who keep trying to define themselves according to what they see happening in the external world. They will jump from one facade to another, always seeking their true nature, yet remaining confused by the faces they see around them which are changing as fast as their own confusion.

Is it any wonder that certain people are admired for being in touch with their own unique self? For those are the ones who have found themselves and do not fit any mold. They are unique and you know you can't be like them because their power is their own. There is no doubt about this.

So, realize the work it takes to find yourself and make the plunge into the inner way. Peel away the false images you present and see yourself as you are - confusing to others and lost. Admitting this will bring you closer to yourself than any other statement you can make.

Do not be afraid to be yourself at the moment of your crisis. Exhibit your true nature as it is known at the time. You will do yourself and others a great service. Say goodbye to trends and, if necessary, become an island in the sea of confusion around you.

Be close to yourself. Separate your true nature from the fads that call you. For in separation there is communion, and in communion there is truth as long as you separate from the external and commune with what is inside.

We say this to help and to give direction. You need to do the work.

24

Allan continued to muse about his work experiences in this place where he was led - a place in the woods far away from all business concerns. It all dealt with the need to address his return to society in a new way, a way he hadn't discovered as yet. And in his recognition of the different roles in which he had served was an answer and it wasn't long before he could see what it was.

He was comfortable and happy solving problems for the people he served. It was natural for him to help people and it made him feel good about himself for he could relieve those problems from their brows by offering solutions that achieved what they wanted to achieve.

But he was not attuned to management, for at those times he was exposed to the inner motivation that moved the company in which he worked, and it was far from what motivated him. And he was not willing to defend the goals of the business.

He saw now that the goals of business were a reflection of the goals of society. They were in fact supportive of a way of life that revolved around making money, growing bigger and standing tall in the face of competitive pressure.

Such business goals kept society growing and prospering

in spite of the service it provided. It was almost a machine that required a certain type of fuel to keep going and he didn't like the choices he had to make to provide the fuel. He could see that the impact of business actions on people was secondary to an internal business sense of achieving growth and power.

Oh yes, the rewards to management were great if they could achieve the inner goals of the business, but it seemed that the people who were served by the business by either products produced or services rendered were a far second to the ultimate goal of business expansion.

And he saw no difference whether the business was of a military nature, an academic nature, a political nature or a business that simply provided products or services. The goals of the business were the goals of society - growth, power, money and a reluctance to allow any potentially good cause to be adopted unless their goals could also be achieved.

Yes, he saw this now, but then he was to close to it. He simply ran away from the machine that wanted him as a supporter or defender of the ways of business and of society - an all consuming machine that ran over the Earth consuming resources in its goal of domination over the competition.

He thanked his lucky stars that the inner urge was there to entice him to run for his life, for he knew that if he was molded into a defender of this way, he would have turned his back on his true nature. He knew he preferred to help people with their own problems and to pay little attention to what made society work.

And in this recognition of where he had been and the forces at work in him that moved him to the place he was now, he realized to whom he was a messenger. It was to the people, not to society.

He had no desire anymore to find ways for society to achieve its goals. His motivation was for people to recognize who they were just as he tried to see the inner impulses moving in himself.

In the whole big picture of life, it was not the money or power that motivated him in the long run. It was a desire to know what was moving him through this life in a way that caused him to choose paths and reject others reaching a point where he now participated in such great drama - a drama bigger than any business, or military force, or government, or any academic institution. And, yes, it was bigger than any man-made religion.

He was now participating in something much larger and he was motivated by this more than he could ever imagine. And the passion plumed up from his gut and lit up his eyes so that he could see what he was all about.

25

We will continue on with some more specific lessons that humans need to hear at this time. We will talk about power today.

Power is the force that is with a person that governs their direction. If their direction is externally directed, then their power is an ego power - one that wishes to glorify the outward manifestation from their head.

This is a self-serving power that can consume everything in the path of what the ego - that conscious part disconnected from the rest - wants in life. There is no consideration for what other people want except for those the ego honors. These are usually close compatible beings who seek the same goals. If these people are supporters of what the ego wants, then the ego will support them. If they are not, the ego will ridicule them for not seeing the value of the chosen path.

The ego's way is self-serving and results in much destruction. And it is this power that governs the majority of modern society. If anything gets in the way of what the power wants, the power increases to overpower any who get in the way. So, humans manifest their destructive nature when they live in service to this consuming power created by a strong ego.

To change to the universal power of love requires great sacrifice for the ego must plunge into the body in order to connect with this power that resides in nature. This power emanates from the creative source.

It is as if the ego of a person is sitting on a high hill defending its territory, while all around it are the voices of the depths singing a song of longing for consciousness. It is this dynamic between the conscious part of a person and the unconscious depths that becomes frightening to the ego.

Yes, even with this all-consuming ego power that is created, it pales in front of the door of the unconscious. And then in fear, the ego power runs away to the place where it can consume the barriers that keep it from its goals - barriers that reside outside.

And as it fights its battles in the external world, it heaps what seems to contradict its way onto the unconscious realm further blocking the passage to its salvation. And fight it will, for the unconscious holds a different power that does not consume but knows how everything relates.

And the power within respects the All, for it is part of that binding force that holds the All in deep communion. It is the love that emanates from the Center.

This is what the ego fears. It fears having to live in a way that is interrelated with the All. It fears the loss of independence and the fulfillment of its goals.

This fear lies just under the surface of the ego-focused human and it increases its power to a raging monster that will fight at all costs to stay above this natural power. The ego knows deep in itself that the natural power can bring the man-made power to its knees begging forgiveness.

And so the ego surrounds itself with those who support its way of destruction and hides from the power within. Yet the power within is patient and unending. It calls for attention in spite of the reluctance of humans to turn within and face themselves.

It calls with love. Not the human love of sexual passion, but a love that keeps all in communion with the ways of evolution that emanates from the Center. It is a power that moves creation to ever increasing pulses of further insight and wonder.

And to be disconnected from this movement creates a sterile garden where nothing can grow save the all consuming needs of the ego-based human for self-gratification, always seeking their self-worth in the external.

Yet, in that same external world that they think they see, are images of what is wrong within themselves. For in the chaos of their cities they see the unwanted parts of themselves in those fellow humans who have become threats to their own safety.

And when they find themselves one night stranded in the streets of derelicts and beggars, they see the dark side of themselves - a side that has been neglected for so long that it seeks them in the terror of the night. These are images of what humans have created for themselves - neglected places that are not in the fold of the All, but wallow inside angry at the ego for disconnecting them from the consciousness that they seek.

It is this terror that must be faced if humans are to regain their true power as participants in the great dance of the cosmos. For only by facing what has been neglected inside of themselves will they be able to relinquish the ego power that now reigns and wrecks havoc on this planet.

Truly, it is a time of great sacrifice. It is a time to sacrifice that self-serving, all consuming power that spreads over the world like a blight, consuming all that stands in the ego's way as it runs away from the love of the Creator seeking its own ends.

26

In our hero's thoughts about his attempts to work in the world of society, Allan had come to the center of his struggles. For ten years he had made a great effort to bring a patterned approach to the world that mattered to humans - the world of business. He had attempted to do this in the military, in the big business of consulting, in academia, and in his own business where he consulted to the Church and to huge corporations that fed the machine of commerce.

And in all his efforts, he had tried to the best of his ability to bring forth a new awareness of a broader picture all in the name of help to people. He had wanted to open their eyes to the connectedness of their own places of work and to bring some pleasantness to their chores. And in all this he had fought the attempts of management to bring him into the fold of their inner group to defend the precepts of this society.

As he passed these ten years, he carried the opposites in him. There was the ego power that sought to be a success and rise to the top of his profession accumulating power and money. He felt this most when he was moved to be honored by his promotions. And he also contained the access to the depths of himself that strove to bring forward the way of the pattern of things in the universe. It was not done in any big way, for there was much more which could have been brought forth, but in a way they would understand for it was in the language of business.

He contained these two sides, the light of activity along with the darkness of the depths just as the Dog Star, in its binary state, whirled the opposites around its own solar system, causing tremendous opposition in the forces present. Yes, he had worked in this tension, for his sourcing was in these very opposites that moved within him.

And in his musings about this ten-year struggle, he could see the hand of the Creator protecting and moving him through the gauntlet that he had to endure. He saw this hand in his whole life bringing him to what he realized was a great experiment.

He had entered this life as a warrior and he fulfilled a great challenge in an endeavor to test the capabilities of the human system to open to a broader world view - a view that would have moved the system into more of the flow of the universe. It would have opened a path closer to the harmonious road of life that led to actions appropriate to the cosmic pattern. In this there was hope that this destructive slide to oblivion could be stopped in a collective way by working with the mechanisms of their life style.

And it was a valiant effort that always led to a brick wall, a wall that prevented the adoption of a broader vision whether the nature of the business was military, political, religious or corporate. All led to dead ends for his effort and he finally had to recognize this and move on to the other option. He knew his effort was to the best of his ability, and he knew it came to no avail.

There was to be no easy, quick answer to the plight of the human species. There was no way to accomplish a movement toward the flow of the cosmos that could move the whole collective of society forward. The only hope was to use the system to do this for it was the one thing that all believed in. And he could see why the effort had failed. The collective Ego of this world could only see through eyes that saw the small world of the physical. In spite of the glow of a more interconnected vision of what that system was involved in, it could only take the pieces that made sense in its own vision, leaving the interconnectedness to die a death on the paper of its creation.

So, Allan left this world to carry out the remaining mission he was sent to perform - to remember who he was and to merge with the Guardians again to bring the messages to the people. It was now up to each person as an individual, each as a mover in their own right of the society in which they lived. Only the movement of each individual could sway the path of this species. Only the openness of one individual, and another, and another could bring the human species to a place where the pattern of the All could be made conscious in such a way as to turn the species away from its destiny of destruction.

And so, in his struggle and turmoil he had seen his mission unfold and he knew the remaining story of the Guardians, and Allyn, and the human species as it played out in this modern era that held so much devastation but also a hopeful seed of a new creation.

27

We will begin with a story today. It is a story that happened long ago. It began in the time when Atlantis sank into the ocean.

As you know, the survivors were distributed into two streams - one to the East, and one to the West. Those who traveled East favored the mind of man, and those who went to the West favored the mystery. We wish to tell the story of the Western trail.

Women of Atlantis who believed in their ability to hold the mystery within them committed to this journey. Women have a natural tendency to be able to hold the mystery in the depths. It was important to do this at this time for the growth of the ego was just beginning and this growth could have jeopardized the future of humankind if the mystery was not held.

And so, the women along with some men followed the treacherous journey to the West and the land that was there. Once they had found land, they began to assess what was needed.

They each committed to following the depths within them in ways that did not seek consciousness fully, for the ego was not developed enough at this time to achieve that. However, they could at least exhibit in building and ritual what was happening within them.

Because of their natural ability to feel the mystery, they became the forerunners of the spiritual leaders of this new land. Many were amazed at their prowess in knowing the rhythms of what was going on in a collective way between themselves and the people who gathered around them. They could sense the fears, the joys, the needs, the wants of these people and what made them sad and what made them happy.

They could also sense the movement of evolution in the cosmic realm and at night they could see this movement painted on their inner eyes. They relayed what they saw to the inhabitants of this new land.

Much was misunderstood as actual happenings in the physical world and not one person was able to make a distinction between the inner world and outer. It just was.

And so, their reality became a combination of the two, but it was played out in outer physical reality. Remember there was not a strong consciousness that could distinguish the differences in the languages of these two realities, so the symbolic significance of the inner way was lacking in their understanding. The eventual building of temples and their art that adorned these buildings simply expressed what the mystery said.

These people were committed to acting out what were the mysterious impulses that surfaced in their psyche. And the psyche longed to be understood, so many legends were created that spoke of this. Many stories that exist in this new land of America have their beginnings straight out of the mystery.

One story we wish to relate. It was on a moonlit night that one woman had a dream. She encountered a man in her dream who was dressed in feathers of a strange bird. It was strange because the feathers transformed in the dream into snakes. And the man glowed with a golden hue when there were feathers, and with a silver hue when there were serpents.

The woman was amazed at this image and told her friends. They looked around the village for this man to see if she could recognize him but no one appeared who looked anything like the man in her dreams.

She was encouraged to dream further and perhaps ask him who he was for it was obvious to everyone that this was a powerful person who could magically transform himself. The scribe of the village began to capture the story as it unfolded so that it could be passed down to other generations.

The woman, after a few days, saw the man again. He told her that he wished to be united with his head. This surprised many of the villagers for he did not appear headless. But, in order to help this powerful man who many were now calling a god, they began to carve various heads in rock to help this god achieve what he wanted. He did not seem to be satisfied with this so they carved bigger heads hoping that if the man was satisfied he would give many riches to the village.

The woman then had another dream where the god walked toward the water and pointed to the East. The villagers talked about this and concluded that the head which he sought was in the land of the rising Sun and so they traveled as far as they could but found no land. They concluded that the god would not be appeased until the head returned across the sea coming from the East.

This dream began many rituals to honor both the East and the Sun, as well as the West and the Moon. Many edifices were built to honor this story and the god who was called Quetzalcoatl was revered all through the evolution of these people.

It was only when the Europeans came to the Americas that the god's wants were fulfilled, yet the visitors came with such devastation. This disillusioned the peoples of this land and they began to doubt the stories that existed in their culture. We tell you people now, that the joining of the head and the serpent is at hand.

28

Allan took some time off from the sessions with the Guardians to assess this path that led to his acceptance of his mission.

He could see the necessity of his early life in the world. It was not only to appraise the capability of the system to broaden its vision, but also to learn the society so he contained their language which could be used by the Guardians to deliver the messages in an updated way.

And he saw his later life as a personal journey into himself, a time to understand who he was. This was accomplished as he consciously went on the inner way, a way that brought him the knowledge of himself and of his mission as he reached the center place. It was a period where he moved away from society's concerns, and followed his own heart.

He could see it all now as a moving drama that led to this time and he marveled at the flow of it. His walks in the woods at this time encouraged him to look at it all, for the woods encompassed the All in its raw power and fertile growth. Nothing was rejected - the dark allowed for fertile beginnings, and the light brought the growth that emanated from the moist, dark Earth.

He saw the new growth hidden in its seed, and how the sunlight opened up the protective covering and allowed the first sprouts to emerge. He listened to the birds as they called out their call to anyone who would hear proclaiming their place on Earth, their voices to be heard. And he moved with them as he contemplated his own journey.

His cabin became a gathering place for all to pay their respects, as animals are wont to do in the opening of one of their own. Skunks came to proclaim their scent, their power to be known as who they are. Wild turkeys paraded their young to him in the knowledge that much could be given. The deer brought with them their gentleness of heart that beat in his chest. And bees circled around the honey, eager to begin work for the Queen.

All came in their own time to surround the energy that was present here as Allan saw the light of day and accepted a broader view than even he was prepared to see, for the center had been reached after years of working with his dreams, and he was moving into the full movement of the cosmos.

As he sat outside his cabin under the stars, he felt the star shine of the Dog Star come upon him and his heart swelled with expectation for it was to be a new world he entered tomorrow full of the knowledge of his life and the way he had been brought. The Owl called to him, a kindred spirit of the dark realms, who could see the wisdom of the depths. And the burrowing animals moved through the dark night, knowing the feel of the dark places on their paws as they climbed out to greet the movement that now would be known in consciousness.

All gathered in expectation as the spirits formed their support around the cabin in the woods, a place where All would be known and the human species would be fed one last attempt. And so Allan returned to his bed, to thank his guides for bringing him to this place, and to begin preparations for his move to the world of mankind, to come as a messenger of long forgotten ways that were the essence of the journey.

He didn't know how the Guardians' messages would be received, but this was not his worry. He was the messenger and he could only follow the way that was presented to him. So, he settled his head on his pillow, and asked for any guidance from his dreams that he needed now, and thanked the Creator for being with him though his long day and night of turmoil in this place of nature that nurtured his way with the power of the pattern.

He closed his eyes with a smile in his heart that flowed to his hands in preparation for the continuance of the messages, instilled with a new power of cooperation and participation that went beyond the expectations he had of himself. For he felt his own power in his center place come alive with the conscious knowledge of his path.

29

Today, we wish to tell humans that much can be learned about themselves by recognizing how they delude themselves.

When you see the world around you, there is much there to see and it can be seen best without the trappings of commercialism. This only tends to draw your focus away from the inner realm and forces your attention on the external.

It is the easy road to see life as an accumulation of needs and wants as expressed onto the material goods that are offered. This is what your society wishes to happen - to allow your own inner needs and desires to be proselytized in favor of the material world.

Much effort is expended in trying to sway your every need away from the spiritual and place it onto the material. You are constantly being bombarded with messages that equate the desires expressed in your innermost feelings with being fulfilled by the material goods that are produced. This is all contributing to your own delusion of the world.

Go to nature and sit surrounded by the sounds that help you focus on the spiritual needs that are coming from within. The sounds of the moving pattern in nature do not confuse the expression of your feelings but contribute to your understanding of what these feelings mean.

The Ancients Come Forth

It is difficult to find a quiet spot in nature without the constant reminder of commercialism, yet if you take the time to get away from the large cities and place your butt on the soil to ground your intention, you will be rewarded with insight.

The quietness of your own inner way can be heard in the movement of nature. She speaks a language that resonates with what comes from your depths.

While you may be fearful of nature, just as you are with your own inner substance, relax into it. Be courageous and see the world for what it is. The cosmic realm moves in a rhythm that excites your own soul and the time you spend listening to your heart and your own rhythms will help you to see where you are.

This is an invitation to strip away the illusion of your reality and begin to become acquainted with your own internal rhythm which is not influenced by the sounds of business. It is a different world that can bring you to a different view of life, a life that is sustaining and fulfilling for it resonates with the pulse of the center.

This natural place calls to you and can show you the path to your own center, the place where your true Self lies in waiting. Do not ignore the messages you hear in the quietness of nature. It is the call of yourself that you hear.

30

Allan awoke to a new morning. He knew he was now ready to partake in his mission and bring the words of the Guardians to the people of planet Earth. And as was his custom, he sat on his couch, opened his notebook to a new sheet of paper and got ready to continue with the lessons the Guardians had been giving him.

He heard the familiar greeting and indicated he was ready to receive more lessons. As the Guardians began to speak, Allan was startled by their words.

"We welcome you, Allan, to your rightful place, a place you have been called to for this whole life you have lived on Earth. We welcome your whole hearted support and will now take time to prepare you for your journey.

"There have been many inroads taken to pave the way for you. This will not be an easy task, but one that will not be daunting either, for there are people today who see the value in the communications you can bring from the spirit world. It is not as foreign to people as it has once been because of the efforts of those who have preceded you.

"The current society calls the avenue to the spirit world 'channeling'. And so you will use your talent to channel to help people as you travel around the country.

"You will have a mobile lifestyle, living in a mobile dwelling called a 'recreational vehicle' and will travel according to our dictates. During your travels, you will also educate people about the inner way. Your introduction to them will be this book that we are writing together.

"So, you will go back to society in quite a different role than you had previously in this lifetime. It will be a role of a teacher. This is natural for you and you need not fear this.

"You have been well prepared by your previous actions in the world to relate to people who are struggling with the power of the ego and its unwillingness to lessen its hold on the person and to join the inner group. This is a very big step for most and you have been through this. Experience is the greatest teacher, so teach from this experience.

"Remember - you are the emissary of the spirit world and as such you will be expected to be open to all experiences along this road of a teacher. Do not hesitate to welcome what is offered along the way, for it may be important in the long run.

"Judge not the motives of the spirits who are with you. They have only one intention - to help the human species accept the path that is harmonious with the evolution of the universe. There can be no other path for the path of the ego-dominated society leads only to destruction for their way of life.

"Be not afraid of those who will label you the agent of the dark forces, for you do have access to this world of the depths and it is not something to deny. Remember only that the dark forces of the depths are a threat to the ego's control and will be labeled by them as evil and tantamount to the Devil's path. This is how they see the threat to their own control of the world and to the worldly power they seek.

"Realize the delusion they are in and simply take their darts of venom as so much venting from a voice that only sees the world as the physical place they inhabit. It is a voice that has lost touch with the broader picture that encompasses the forces you have dealt with and communed with. It is their way to see this way, and nothing other than your teachings about the inner way should be hurled in their direction. It is only when they have entered into themselves that they will cure their own delusion. Realize this and go on with your mission."

31

We will continue on with more ways people attempt to ignore the depths.

There are stories about the depths that are perpetuated in today's society. People say it is a dark place of evil. If that is true, then it seems humans are walking around as explosions waiting to happen.

Yes, this is true but not in the way that is perpetuated. If the depth material in a person is not addressed, then this material takes on a sinister character and tries to be seen by projecting itself upon the world. It is only through the consciousness of each individual that the depth material can be seen for what it is - the parts of themselves that need to be known and dealt with.

Left to its own devices, the energy becomes dark and negative and has devastating effects on not only the person who owns the material but also on society as a whole as the dark portions are released without knowledge. And it is the fear of knowing this material that keeps it negative.

So, what do the stories accomplish? The stories give credence to the fears of the ego and allow the inner realm to be put into the negative image that the ego desires with devastating results on the person and society in general.

There is also the story about the depths as a place of previous evolution. Here, because the depths are considered to be in a downward direction, the story holds that plunging below is reverting to some primitive savage part of themselves that will reverse their flow of evolution, causing the person to take on primitive, unacceptable behavior.

While it is true that the depths do contain primitive parts of the person, it is primitive because it has not been brought to consciousness and allowed to mature. Rather than dealing with these parts of themselves, they are relegated to the dark places where no clarity can be achieved.

As a result of ignoring these parts, the person will, in fact, exhibit behavior that takes on these primitive characteristics in their full negative condition. So, rather than the person making the effort to bring forth these primitive aspects and make them helpful parts of themselves, they allow these primitive, savage feelings to come over them unexpectedly to wreck havoc upon their well-ordered society. It is only in this way that these aspects of themselves can be made known, albeit with negative consequences.

In both of these cases of ignoring the depths, the devastation of these parts of themselves is allowed free reign to cloud not only their own behavior but also to negatively influence the collective state of their society. It is only by facing the demons, as they are called today, that consciousness can be raised to a point where the human is fully realized to their own potential.

Remember, that when parts of themselves have been ignored, then the negative aspect is released with full power to gain the attention of the person and society in the only way that is left to this realm, for the inner depths always seek the light of day whether they are invited or not. The courageous action of some humans to meet the inner realm with a clear eye lessens the negative aspects in society and contributes to further evolution.

We tell you this to show you how your stories only tell the truth you wish to see so that you are protected from what you do not wish to address. The ego, in its shallow way of seeing the world, wants to stay shallow for the plunge into the depths is a threat to the ego's own power and desire to achieve its own aims.

It fully knows in a consciousness long forgotten that the ego is not to stand alone but be a participant in the whole of the person. This requires a recognition that all is related and must work together to achieve the evolution of the spiritual. The human contains all that is needed to achieve this evolution and only requires the courage to face the dark portions, those parts of themselves that have not been brought to consciousness.

Delusion is a safety net that keeps the human lost. We cannot emphasize this enough. Look at your own stories that keep you from going inside yourself. What do these stories say about your own fears, and what truth lies between the lines in the story?

Can staying in a shallow, confused, delusional manner contribute to anything but to further the confusion? Are you so afraid of the rest of you that you choose a life of misery?

Awaken your heart and listen to what it says to you in the quiet of your bedroom as you lie there on your bed and contemplate the day. Is this a life that is fulfilling to you? Where does your heart call you? By all means, connect to this loving voice within you and listen. Your salvation is in your hands. The door closes on your side, and opens also. It is your choice.

32

The Guardians continued with their message to Allan.

"You will also meet groups of people who are reluctant to do the work but want to reach this higher vision through exaltation of the ego's place. They will try to emulate what you do and say through the actions solely of the ego, while maintaining control over their personhood.

"You will see this for what it is. An attempt at a quick fix and you will have to encourage them to see what they are doing. For only when the ego joins the inner group will they be opened to the broader vision that is necessary.

"Do not berate them but turn them to the inner way by helping them with their dreams which they are reluctant to see. Relate your own fear and turmoil as you fought the inroads of the depths. Tell them your story of agony as you saw what was needed.

"You will do little good telling them how wrong they are, for this just reinforces the ego in its fight to maintain control. Rather, teach them with your own experiences showing the egos that you meet that it is painful but not impossible. This is the best approach.

"Listen to their dreams and tell them what you see even if it is not accepted. Do not tell them what they want to hear -

tell them what you see in their dreams. Many will walk away but they will remember that someone spoke the truth to them and this glimmer of truth may be the seed that gives them the impetus to take the plunge into themselves.

"We tell you all this to prepare you. Remember these words well for they will help you in your journeys."

And with these words, the Guardians closed the session.

33

There are many ways to relate to the cosmic realm. There is the inner way of working with dreams that bring you closer to who you are. As you participate in this work, you are participating in the cosmic flow as it is translated into your local personal realm.

Each thing, person, living entity has this localized rendition of the force that binds the All together - the love force. In your personal realm, the force manifests in a way that draws your consciousness into a deeper place to see all that moves there.

It is necessary to go to these deep places within in order to be in communion with the movement of the cosmos. Staying above this level leads to independence and your movement is not in harmony with the larger movement of the universe.

You are like a small unruly boy who wishes to get what he wants and ignores the advice of his elders. In ignoring this advice, the human turns his or her back on the great movement of the Creator.

And it is no good to follow the advice of those you respect in the outer world unless that person knows what is going on in your depths. This is very difficult for another person to know. If another person tells you to do this or that and has not taken the time to know who you are and what unique movement you contain, it is best to ignore this human advice.

Be aware of this in all your dealings with other people. It is only those who help you to know what goes on in your depths who can be of assistance.

Each person has their own movement. Each person is an expression of how they participate in the cosmic dance. Forcing a way does not help you nor the universe. It tears a path through the flow creating holes in the love force that tries to move you in harmony.

So, it is necessary to go within yourself to heal these past endeavors to force some conclusion. This is painful for you have to face the bitter truth of yourself. It is bitter because it may contradict what you have built up in your thoughts about who you are and what you are desirous of doing in this life.

All these thoughts have been influenced by others who have told you what they see in you. These views of you are based on their own unfulfilled journeys and they are trying to live that through you. Only those who have worked on their own depths can truly see who you are. That is why we say that you should only follow advice from those who know the importance of the depths.

Throughout humans' time on Earth, the depths have been looked on with great fear and images of evil creatures invented by humans have been associated with this place. Realize that the cosmic flow has many deep places where much creativity is happening all the time. While the deep places are frightening because they reveal the bitter truth, it is better to see this rather than go through life unaware of what is happening in the cosmic realm.

People are not made to be independent from the All. They

are intimate portions of the great All and as such must participate fully to give back to the Creator all that has been received, even when the person is unaware of what that is.

Flow goes within the human and then flows back to the Center. It is this movement that is essential for evolution. When you refuse to listen to the messages in your depths, you do not participate in the flow and you establish holes in the fabric of the universe. This hinders the movement and you become like a blight on the creative impulse that is flowing constantly.

It is not necessary to go through life this way, feeling disconnected and alone. The holes you create through your lack of participation feel devastating as if you had no purpose in living and no idea who you are and why you are here.

This is not necessary. Have courage and go within. Listen to your dreams. Share them with others who know this realm and receive help with caution, for only you can know what the messages are. They are about you, no one else.

You will know who shares the truth with you for it will bother you. The bitter truth is not pleasant. Use this bothersome feeling as your guide. If you sense that your feelings tell you that this person leads you astray, away from the bitter truth, run away.

Each time you will learn the language of the inner way, gaining insight as you go. For no one can really tell you who you are except your inner messages.

Pay attention and live. It is only in this way that you will begin to move toward your center Self, that person that is you. It is only then that you will make movements that blend with the All and you will no longer be alone.

34

Allan contemplated the Guardians' message realizing that he faced a society of those who would not accept the message at all for they saw it as an evil threat to their way of life. And he also knew there would be those who accepted the channeling as a new age good thing, but would seek a quicker route in the manner of one who had done the work. They would attempt to put on the clothes of the inner traveler without making any effort to delve into themselves.

He had met these people already - people who saw dreams as prophesies about everyone else except themselves. Yes, these would be the hardest nuts to crack for their protection was clothed in a different version of ego control - a control which identified the rest of society as wrong. They considered themselves dissociated from this larger group. They called the ego bad and convinced themselves that the ego had left them, and now claimed they were their Higher Self, all within the blink of an eye.

Yes, he thought, he had met those egos and he realized the truth of what the Guardians spoke - only the truth of their dreams would jolt them into what was needed - the plunge into themselves. How delusional the world was and how delusion was protected and fought for.

Allan mused about these things and realized his role would not be acceptable to many, but he was a warrior and used to the ways of his road. Many would attempt to thwart him on his path even as he taught the inner way and his warrior courage to go on would serve him well.

As he contemplated his own sense of himself, Allan realized the difference between a fighter and a warrior. He saw the fighter fighting against external forces that would impede his progress. This was not who he was, for in his heart there remained no anger at the way of the world.

No. His heart beat with his own sense of who he was and what he was all about. For the sense of himself rose from his depths and this was his authority for himself. There was no need to fight his detractors, for they did not impede his path. His journey was motivated from within and this they could not touch.

He knew now the heart of the warrior - one who taught what was within, and was confident in his path for it came from within his soul. It was not the fighter's bravado, but the warrior's heart that had the courage to be himself. And in this, Allan knew that in spite of roadblocks put up in front of him, he would walk his talk and feel the strength of what was given him.

35

We will continue with a lesson about the way of dreams. Dreams are communications from the inner realm. Dreams refer to the journey of a person, telling them what things they need to bring to consciousness about themselves. These things could be their fear of giving up a way of life that does not suit them; they could be about learning various aspects of their personality; they could be advising the person about choosing a new path or giving up an old way.

All these messages deal with bringing to consciousness the flow that is happening in their inner depths. This is to bring them to an awareness of what is going on with them and to help them move to a path that is in harmony with the depths of themselves.

The dreams will also bring back all the parts of themselves that have been discarded into the unconscious, if such parts need to be known again. Many of the messages will reveal things that are not pleasing to the person's ego for they will be about the truth.

The ego may be uncomfortable seeing the truth of its personhood and will try to hide from this. Therefore, in this day and age, it is best to write the dreams down as honestly as you can record them. If you do not, the ego will attempt to discount the dream images that are not pleasant to it by bringing doubt and may even try to distort the images to make them more pleasant.

Also, there is a great tendency today to see the dreams as messages about other people - friends, family. This is another way the ego attempts to push the truth away from the person and onto someone else. We caution all those who are working with dreams to remember that these are messages about themselves in order to awaken them to who they are.

Even dreams that seem to come true for another person are still about your own journey. For example, you may dream about a friend who has had an accident. You do not tell the person about it and the person does in fact have the accident. You think you are having dreams about others and you are at fault for not warning them. We tell you that your only fault is not addressing the message of the dream for yourself.

The friend represents some aspect of you that is in danger because you have not addressed this part of yourself. By thinking it is about the friend in the external world, you project the accident out into the world and, if your negative energy is very strong, this could bring devastating results into manifestation. This is just like not being aware of your darker qualities and they then find expression in a negative way.

Many problems in today's society stem from this dynamic of not being aware of your own qualities and then they are projected on society to wreck havoc. If each person would own their dreams, much would be accomplished in the world of your making.

Dreams are very powerful and need to be understood. They are not mere rambles of the mind nor are they superfluous to your life. They are your guide in how to live your life.

The language is symbolic and ancient. The language incorporates symbols from your own society. It can be learned just as you learn a foreign language, yet experience is the best teacher.

Talk to others about your dreams and see what they see. Look up the meaning of the symbols that are in your dreams, always remembering that the context of the dream colors the meaning of the symbols.

The entire dream is interrelated, as well as the dreams of one night's dreaming. The dream, and the symbols and actions in the dream all are trying to get a message to you about your life and who you are.

This is vital work. There is much knowledge of dreams in your society that are held by a few people. Seek them out. Learn by experience. Make the time in your life for these messages.

Remember that this is the best avenue for you to discover who you are and how to begin to live in harmony with the cosmic flow. You will be nourished and fulfilled if you will just begin.

Ignoring a dream is the same as ignoring a message from the All. This only hurts you.

And if you say you never dream, realize this means that you have closed the door to your inner life and put locks on the door. These locks are blocks you put up to protect yourself from the messages. You have great fear and must begin to address why you fear opening to the cosmic realm.

Do you feel insignificant? Do you fear you will be devoured? Do you fear a life change that will disrupt your stagnant life? All these questions and more need to be addressed.

Be courageous and step up to your true nature. You are

spiritual beings on a journey of discovery. Listen to what is said and grow in consciousness. It is time to wake up.

36

Allan had been through a major inner struggle that had gotten him in touch with his mission which he had finally accepted and he had found the inner strength to go on. Also, the Guardians had begun to advise him on ways he could deal with others as he became more of a public figure.

Allan had digested all of this and was in a good place to become an emissary for the inner way. This would see him well into the future of his role as the messenger of the Guardians. Now it was time to focus on the way that he would begin - a way of being that was unusual for the ego, for it was a change in ego makeup.

The ego was initially tied to the physical realm. If no changes happened in a person's life, the ego died when the body died. This had now changed in Allan. The inner way had propelled the ego to the center place where the ego was an emissary to the cosmic flow. This also changed the body from being less of a physical entity to one of more spiritual energy. Allan now had to deal with this change.

On the morning following his own understanding of the strength that resided in his center, he continued on with the Guardians who provided further messages to him as preparation for his work. After their usual greeting, the

Guardians began to teach him about the body as a physical manifestation of the spiritual energy that was contained within the center place. He was told about the spiritual essence that surrounded him and most humans who had any inkling about this spiritual force.

If a person was so relegated to the physical, this spiritual body had little power and acted as an insignificant force in their lives. It appeared as a very faint aura surrounding them, thin and weak and unconnected to what was manifest, for their ego had sole control over the way they presented themselves. So, the connection was very tenuous to the spiritual force that surrounded them and thus they were not influenced by it.

Once a person reached the center place and dealt with the issues they needed to address, primarily around the particular mission they had been given, the spiritual essence of the person took on the prominent role in their lives. This is what had happened to Allan and there were instructions he had to receive, for the way he went around in the world would dramatically change. In other words, he would relate differently to other people. Allan had not seen this yet, but he would notice it in the interactions he had with others. And so the Guardians began the instruction.

He was told about his sense of time. Allan had always been very quick on his feet and was able to discern the intricacies of the biggest problem, offering solutions before many had a chance to digest what the problem was.

His sense of time was very detailed. Each second was a major event in his life. People before had told him that time was a major ruler in his life and he knew this as he saw his own fleet of foot. There were times when he was taken quickly to a conclusion and he became frustrated by the slowness of another to follow where he was going. He liked to go fast and preferred that people he met also move at the same pace. This was not the time sense of the center place and Allan would now be moving into a flow of time that bordered on the eternal.

This was the biggest change Allan would experience and he was told to be aware of this, for the consequence could be great if he did not see what was really going on. For once a person was governed by time, and then all of a sudden their speed of processing changed, there was a tendency for the person to doubt their abilities in spite of a keener insight that was present.

The person became like the center of a tornado, calmly looking at the turmoil surrounding them. They were able to discern the flow of human activity, yet at the same time they moved according to a higher power that brought them much attentiveness without the show of participation in the human flow. It was like a still point in a storm that could be in the turmoil yet be unaffected.

This was the basis of Allan's strength now but he was unaware of the implication. And so the Guardians began a series of lessons to acquaint him with this new place before he began to think that he had lost his ability to contribute what he thought were his best attributes to society - his quick answers to the problems that had beset the human race.

It was no longer his role to do this. This came as a shock to Allan for he always thought he could solve most problems in short order and it was this capability that made him a contributing member of society. Now in his new role, he had to see that it was not for him to solve other people's problems, but only to turn them toward their journey to their own center place and thus allow them to discover for themselves what their own mission was, and what issues they had to address.

And so, as gently as possible, Allan was instructed in the way of the center place and was shown how this new time sense would assist him with the patience he needed to work with others as they plunged into the difficult journey of discovering who they were.

37

We will talk today about the ability of the inner way to help a person begin to know that there is more to life than the external world they see around them. Once the human has opened the door to the depths, the inner way will begin to teach them how their current activity in the world is helping or hindering their journey.

It is not surprising that the inner way will usually inform the person that their actions are not in accord with their inner Self. This is because they have been shut off from the messages that come from within.

So, there are many bitter pills to swallow at first. Not that there are many painful messages, but the ego will not be very pleased to accept what is offered, and the messages will have to be repeated many times until the ego begins to wake up to reality and see what is going on.

This process may take a long time depending upon how strongly the ego is convinced that its way is the proper way. The ego must deflate itself in order to hear what is being said.

There is also, on the ego's side, help in the fact that the ego does not know the language yet and can read many signs into the messages that its path is a good one. This is why we say that a new initiate to the inner way needs to seek help in understanding the language of dreams. This will help ameliorate the tendency to misinterpret the messages being given. Once the ego is willing to lessen its control over the person's activity, then the process will gain momentum and the path on the inner way has begun in earnest.

There is a slight tendency today to think that the ego is a culprit and needs to die, in a literal sense. We tell you that this is false thinking, for the ego is a necessary part of the human. There is no life without the ego for it is the key to consciousness. Thinking that the ego must die, in a literal sense, will only create a delusion inside and inhibit full consciousness.

Yes, the ego needs to die symbolically, for this is the deflation of the ego into a form that is willing to participate in the inner way. So, do not fall into this trap and think the work on the ego can be dispensed with in a quick death. Realize the delusion contained in this thinking. It is simply an attempt to get out of the painful work of dealing with ego ways and relinquishing control and power so that the ego begins to listen to the wisdom that emanates from within.

This is a necessary process that cannot be bypassed. It is painful, yes, to see the control the ego has over a person's life and how in its search for external gratification it is confusing the journey of the individual. This must be seen so that the wisdom of the inner way can be appreciated.

So, it is best to get on with the work and accept the pain. The sooner this is done, the sooner will be your rewards. Remember, there is much that has been submerged into the unconscious that the ego did not like. Now, each layer must be peeled back to allow the person to reclaim the parts of himself or herself that have been discarded for the sake of achieving external rewards.

We know that this society wants quick solutions, but the process of ego deflation is generally a slow one. The control of the ego is strong, and it must be shown the consequences of this control.

Many battles will be fought by the ego to keep the truth from being seen, but in the end, the ego will finally realize that it is not alone. This is a major step forward and will contribute to the journey of the individual for it will allow more of the messages to be received and the actual work of finding the person's appropriate path will commence.

So, realize the pain that will be experienced at first. Your ego is strong enough to take it and will become a better participant because of it.

And realize it will take time, for what is time but the opportunity to evolve. It would do no good to accept all the painful lessons at one time, for the lessons would be lost in a sea of confusion.

Take the time each day to see how the ego has duped your vision and let the ego see the benefit of knowing reality as it is. In this way, there will be no reversals of the process for the ego has been instructed in easy steps that it can handle and will not very soon forget the process.

The process is necessary. Take the time. For in learning patience, you will be better able to love yourself. Try to speed it up, and you will grow frustrated at yourself and begin to despise what you see and loathe the parts of yourself that can be your best allies.

Be patient with yourself and realize much can be accomplished in short periods once you have spent time with the current condition of your inner self and the control the ego has over the course of your life. Time is not an enemy but an opportunity to discover who you really are.

38

Allan entered into a period where his new found behavior was couched in the realization that he was leaving a mental attitude that had been forced upon him while he was still young - an attitude that said he was to excel in spite of others. This was an overriding precept of society that was the impetus to the great competition felt in people's work and in their everyday living for this precept permeated all beliefs.

Striving to be better was built on the premise that better was equated with higher standing within the society's hierarchy. Thus, better was promotion, better was a contribution that enabled society to move ahead over obstacles, and better was more money and power.

As one person advanced over others, they felt rewarded while others felt left out and demeaned. This was considered consistent with the goals of society for this lowly feeling created in others was looked on as an incentive for them to crawl out of their lowly existence. And so in spite of the turmoil created in society by the incentive to rise up in stature through job and position, it was ignored as just something that would eventually work out as others brought themselves to higher positions.

Yet, all could not rise to higher levels as the population

grew. The number of jobs with higher prestige could not keep up with the idea that all could crawl out of lowly places. This intensified the feeling in many who were getting out of the bottom rung that each higher place needed to be held onto with a fierce grip, holding onto the capabilities that brought them to this place that was more comfortable.

To act in a way that did not show off a person's talent for solving problems, for this was the task of the higher echelons, was tantamount to a fall back and this is what Allan was feeling. For in pointing the way for people to find the answers within, he took himself out of the role of giving answers to them and this lessened his own self-esteem because it was based on showing that he was worthy to lead a society forward over the obstacles that presented themselves.

This led to a great chaos within him. While he saw the wisdom of his role as a messenger, bringing a message that required all to go within and find their god-given talents, he could not deal successfully with losing his powers within society as one who could solve problems.

It was difficult to sit back and say that the problems facing the world of his culture could be dealt with by everyone going within themselves and learning what was there. He had to face what was the source of his self-esteem and this was far more difficult for him than anything else that he could contemplate for he thought he knew what self-esteem was. He thought it was the feeling that came from a job well done, a problem solved and the resultant admiration from others.

Yet, he could see a falseness in this. He had long been suspicious of the admiration offered, for there was always that underlying jealousy among co-workers that said to him why wasn't I included in your problem-solving, why couldn't I assist you. Why was it you who could see what needed to be done faster than I could?

And so deep down, he was suspicious of the congratulations and admiration offered, yet he knew that if he was successful in offering solutions that others could not, management would have no other choice than to promote him to a higher place of prestige. And he saw for the first time that he was, in his acceptance of this way of society, helping to promote the ego's inflation of its own power to control. If he was known as a problem solver, his capability for intelligent decision making was being elevated to a place where he could direct the way that solutions would be achieved and this gave him great power over the movement of society.

He recognized this in his past life. He saw how the solution he proposed was more important than taking time to assess it against other factors that would ameliorate his solution in favor of more appropriate ways - ways where the ramifications of the solution would be seen. But this required time to listen to your heart and he was reluctant to take the time. Time was critical for it determined when a solution was either looked on as good, or not as good depending upon the situation at hand which was always changing.

Oh yes, he had learned their lessons well. Even when he began to see the pattern moving in all things, he was reluctant to stay with it for fear that the solution he had recommended would not be as glorious as the way he explained it.

39

There are many ways humans attempt to block the inroads of the depths. They involve themselves with many activities to fill their lives. This allows them a semblance of feeling that they are accomplishing things and provides them an escape from being with themselves.

They know that deep inside them lie the answers to the questions of their life but they also know they will have to get into their feelings of being inadequate, of being alone, and of being lost. This is not pleasant and the activities keep them from experiencing these feelings.

Yet, the feelings are still there and in moments of quiet they still feel them. So, they add more activities to their lives to keep these inroads of the depths from being felt. This is, of course, delusional behavior for they are deluding themselves into thinking that all is well as long as they are active and accomplishing things.

Many others are willing to go into the depths, but they go into the depths with their own agenda. They want something and they go into meditation or into their intuition to seek confirmation that what they want is appropriate to the inner way. It is a false picture they receive, for they have not learned the difference between their own mind games and the voice of the inner depths.

They see meditation and intuition as better than listening to dreams because they get the answers they want, yet they have no experience in discerning the difference between their ego's voice and the voice of their inner self. They do not wish to expend the effort to learn the language of dreams. This is unfortunate for again they delude themselves into thinking that what they desire is appropriate to the inner way.

In general, without knowledge of the inner way gained through hard work with its language, the ego is not ameliorated by the forces that lie within. And so, it is quite content to interpret the inner voices in its own way.

Can there be any doubt that the ego is threatened by the task of learning the language of the inner way? Which way do you suppose is appealing to your own ego - to go the way it wants or to condescend to listen to a more powerful voice that has the wisdom of the ancients behind it?

Much effort then is expended by humans to keep hold of what they view as a rational path in life, one that is not threatening to their wants and needs and one that does not have to listen to alternatives that come from a place that is not honored by this current society. It is a safer route for the ego and one that can ignore the ways that threaten the rational basis of the head.

Finally, there are attempts by humans who know the wisdom of the inner way to close off the avenues to their own salvation because of fear. They have evolved to a point where they know the virtues of the inner voice, yet fear it because they know it will lead them to a place where the external reality they believe in will crumble and they know not about what will replace it.

These are people who are on the edge of full awakening.

They are suspect of the way that society has garnered but realize that they are reluctant to be seen as different for it will take great strength to keep on their inner path. They fear the broad vision that comes with this.

It is difficult to ignore things when a person can see for miles like an eagle and can discern the differences in one way and another. They fear this view because it is too clear.

There is no delusion in this broader view and they have grown accustomed to the protection that delusion affords. Now, to take away this protection feels as if they are totally exposed to the cosmic elements and this brings fear that they will be devoured as something small and insignificant. It is a feeling of being "out there" ready to be set upon by voracious monsters who will push them this way or that and who will hurt them with their great power. So, their childish ways keep them holding on to the safety blanket of delusion for fear of knowing too much.

To these people, we can only say that you are universal creatures who have been created by the cosmic realm you fear. It is your true nature that has you cowering in the corner of your soul fearing to know the greatness of your own spiritual nature.

Yes, it is different from what your ego views because your ego has a small view of who you are. Yet, you are on the edge, are you not? And you hear the call to take the plunge into your true Self. If it is a spiritual plunge, do you really think that it will bring you to disaster?

We know the next step is not easy. Have courage and believe in your own worth. Face it with a dignity that becomes you.

40

Allan thought about how he had sought after self-worth, confusing his own mind with what he thought was good, or with what he thought he had discarded long ago. Now he faced a hurdle as these societal precepts returned to question this new path that led to a different kind of self-esteem, one based on a path that lessened the ego's role. As a messenger, he could not take credit for the message but simply help the manifestation of each person's soul.

He stopped in his musings and saw the greatness in this, yet it paled in society's view of what was worthwhile. There was a transference from an ego's view of self-worth to a self-worth attributed to inner ways that seemed to sacrifice ego power for the good of all. This stuck in his throat constricting his usual way of opening, for it was tantamount to giving away the store.

He stopped in front of a mirror and looked at his face. Yes, he said, in my own face I see the truth for in the eyes I still see the desire to be seen as powerful in my mental activity. He looked at his ears and could see the lessened recognition he gave to the capability to listen. He saw his mouth as the proclaimer of answers rather than the bearer of truth. And he smelled his own scent as a combination of ego gratification along with the stench of being a facilitator of others' capabilities.

He felt his constricted throat and knew where the problem lay. He was fully exposed, torn between two senses of value: one from society and one from the inner world and he did not like what he saw. He knew the wisdom of the inner way and his own sense of worth was now in question. What had brought him prestige in the past was now in jeopardy, and what would be the inner truth of his own worth was not yet seated in him to provide a compensatory feeling of goodness.

He left the mirror to go face nature outside his cabin in a state that humbled him greatly. In all this journey to the center place, he had fought in the hidden reaches of himself with self-esteem, and now it faced him with the power of fire, ready to transform him finally into something that he feared in a way, for he would lose something that had motivated him for years.

41

Today we will move on to the phase of the inner way after the ego has agreed that it does not know the way by itself. The ego is now willing to join the inner group of people that appear in dream and meditation.

These inner people are all the aspects of the person's center Self. They are the faces or facets of the diamond that resides in the center place.

Each face that is presented has the ability to teach the person about an aspect of themselves. It is the task that must be addressed next if the person is to move toward a place where the ego feels comfortable working with all these inner figures.

This can be a very rewarding time for the person, as the unconscious aspects of themselves begin to emerge and are related to the day-to-day activity of the individual. There may be conservative elements that surface which are inhibiting the person from seeing the world as it really is. There may be wisdom guides who appear who ask pointed questions about what the person is doing. There may be guidance offered to show how the inner group can assist the person in various activities that they are undertaking.

And the ego can be seen in the dream as the one who is

refusing the gifts offered, or is frightened of some aspect of themselves, or is always saying something to which the inner figures do not pay attention. These actions are all trying to teach the ego how it relates to the movement that is going on within.

Everything in the dream is instruction to let the ego know how it is relating each day, helping it to move into a closer relationship with the inner movement. In other words, it is helping the ego to find its rightful place in the group as not only a student but an active part that can assist in the manifestation of the center Self.

Each lesson that is learned will bring the ego and consciousness closer to understanding who the person is. This is a period of cleansing of old ways and a period of transformation. For in the movement of the inner way, there is always the process of change, and the changes can be small steps or big leaps for that is the way of evolution.

As the person evolves out of this inner path, the transformation begins. Some movement may seem difficult at first, but in the long run the great benefits will be seen. For can there be any greater gift than to know who you are? Have you not been searching all your life for this knowledge?

Is it any wonder that your eyes will look at yourself in the mirror with a new light once you begin to discover all the facets of your Self? For you are a wondrous being who can look at yourself with the eye of consciousness and say, Aha! I've found out something else about myself.

In this process of awakening, the person will experience a great euphoria at times, as the vision of the outer world begins to be affected by the inner way. There is no greater feeling that can happen than when a person has a new awareness of themselves.

This awareness will continue to grow as the group along with the ego begins to approach the center place. For in the process, the ego begins to belong to a group that has its own best interests at heart. Rather than being alone, the ego can now see that it is participating in a great dance that has significant purpose - the awakening of the person into consciousness.

Much can be accomplished in this place, and much of the outer activity will be put into question. The person will begin to question how they spend their waking times and will be given many images to help them move to an outer reality that is more attuned to who they are.

This requires great openness on the part of the person, and the rewards are great. Do not expect too much at first, for the ego is still learning that it is not alone. But as it sees the benefits and finds its role in the group, much can be accomplished toward creating an environment that is fulfilling and nurturing to the individual.

A great implosion can occur as the ego deflates and finds out that its role is significant but in a way that has never been known. It is at this point that the ego sees with new eyes the movement of the inner realm and realizes how lost it has been. Then the work can commence in earnest, seeking the center place where new rewards are offered, yet new fears are realized.

42

Allan's torment about his self-esteem and the source of it was with him throughout the day as he spent most of his time walking in the woods and feeling the presence of nature around him.

He could see how his self-gratification was coming from putting himself first, showing his prowess in the world, trying to elicit admiration. He knew this behavior was obsessive in a way for it was encouraged by all he saw in the world. He also knew it was ego gratification - nothing more, nothing less.

This realization stared him in the face all day and there was little help in the woods except seeing that the natural world did not partake in its own glory. Here self-esteem was found in the flow of nature moving through each one, being part of the cosmic movement. And while in seeing this there was an awareness for him, he was too distraught at seeing the truth of his own behavior to focus on a replacement motivation that would have brought him quickly to see what was moving through him. He was in the flow of nature but was distracted from it by his own concern for his self-worth.

And so he moved in spurts, finding more realization in the releasing of a form of gratification that he knew wasn't appropriate, rather than realizing that he was already a part of the cosmic movement.

43

We will continue with some more concerns we have about the way this society is facing its role in the world.

There are many ways to ignore the opportunities that are presented by the universe. One of these ways is to engross oneself in the making of money.

Making a living is not bad in itself. People must live and barter for what they need since the complexity of society hinders one from living simply off the land, although some do this. Therefore, some form of value must be exchanged in order to receive the sustenance one needs.

However, to make money for the accumulation of power and glory does not fit in with its relational aspects. It is an exchange mechanism that incorporates many attributes of the old barter and trade ways.

This is the positive way to deal with money, as a means of exchanging what you have to contribute to society, giving back to the group that you live in. In other words, money is to be in constant motion, exchanging goods and services for the welfare of the people.

Yet, today money is hoarded and kept for power and influence over others. This is not its purpose. Nor is it helpful to accumulate money to counteract the fear of change, or to address the concerns one has for the future. In this way, people avoid dealing with situations that are presented to help them move spiritually. They use money as a protection against change and necessary movement in their growth.

Today, this purpose for money is taking on the predominant meaning. And so it is demeaned and loses value, and thus more money is needed for protection. This use of money today to protect people from change blocks the power of the universe to assist the movement of evolution.

Do not think that this is said lightly. It is a force that inhibits positive change and so the evolutionary forces go within the dark realm and begin to wreck havoc upon society in the place where nothing can be dealt with - in the dark corners where it is hidden from view.

It is important to recognize where change is now coming from. It is feared as a dark force that threatens the underpinnings of what people value - a stable lifestyle supported by money and things that money can buy. Thus, what people want today is protection provided by money that is not in movement, rather than money as an exchange vehicle.

And when they need more wealth, they find ways to borrow the money that they wished they had to surround themselves with things that give the semblance of a good stable lifestyle. It is all a facade. Money has been put into a place that cannot be supported for it will naturally fight the desire to keep it stable, for it is a moving invention designed to provide for the exchange of goods and services.

This will always be money's positive role. When it is abused, then it takes on a sinister role that is unpredictable. This is what is happening today. And it hinders the spiritual growth of people.

It is necessary for each person to see how money has become a hindrance in their lives - a hindrance to what is being called for in their own depths. Both money and spiritual movement can go hand in hand, but it requires money to be fluid, not held static by societal pressures. Such is the way of it.

We say this for all those who seek protection in money. It is not the protection you think it is for it will fail all attempts at stability. It is not in the nature of money to be stable and all attempts at relying on money for protection will fail.

Leave money in its fluid state and be open to change. This is one way to balance the forces that are now at work. We caution all to look at money for what it is in their life and recognize the hazards that are there.

44

Allan's turmoil continued on through the night and his dreams offered no help for he was already there and there was nothing more his dreams could say to move him further. The only ingredient needed was to realize where he was and he was not willing to let go of his own meanderings in the pit of depression where he found himself.

This was not unusual, for in any realization of something detrimental that was working in a person, a period of mourning had to be experienced as the old was dealt with and all the feelings of inadequacy at not dealing with it previously were felt. For in most people, realizing their own lack of recognition in the past further propelled them into their lack of self-esteem, which their self-gratification could never fill. So, wallowing in the pit of themselves only helped to show them how shallow their self-esteem was as they saw how they had to continue to seek admiration from the outside world to constantly build up the facade of self worthiness.

So, Allan's despair was needed to further clarify how inadequate the self-esteem was that he thought he received, for it was short lived and needed constant feeding in the competitive nature of society. Through the night, he argued with himself wondering why he sought this way of feeling good about himself, only to find the answers in his own lack of recognition of who he was.

And he detested this lack of recognition of what he had been doing all his life to seek self-worth, for it brought up his lack of consciousness about himself and his behavior. It was even more painful because he had followed the inner way for so long and had never seen this, even though he remembered all the ways his dreams had tried to bring him to this point.

How he squirmed out of it, he thought. It was at the center of his core, buried deep in other issues that he had to address. It was always there, in the far reaches of consciousness, just at the fringe so that he was drawn toward it even as he rejected attempts at addressing it before this time.

Now the time was perfect, for his time frame was changed and he now had the fortitude to stay with the process, even though it pained him greatly to see himself in such a strong light that showed all his flaws in full detail. There will never be any more pain than when you have to stand in front of your mirror and look intently at all the flaws in your face. Yet, in the flaws were the cracks in the armor that could bring forth the new creation.

45

Today's lesson is about the way that people hide from their true nature in spite of receiving messages that entice them to embark on a new path.

Taking on a new path in life is quite difficult. There are many things that keep the person locked into a way that inhibits their growth. The need for money is one of the most powerful.

The amount of money that a person needs to sustain them is very variable depending upon the amount of possessions they have acquired. This is recognized today by a trend by some people to simplify their lives, which amounts to assessing their possessions. It is seeing which possessions are really needed and which ones have taken on the role of ruler in their lives making them slaves to the possession, and living a life that is spent constantly trying to earn the money needed to keep them.

This is very destructive for such possessions take over one's total attention and keep the person acting in ways that are totally opposed to their very nature. And without addressing these possessions, there is no way out of the predicament.

It is very difficult to concentrate on what the inner life is saying when the person's waking hours are consumed with the worries of making enough money to uphold their life style built around these ruler possessions. This is why we say that dreams become the only avenue to hearing what the inner life is saying because it is the only voice that can be heard within the frenzy of activity that rules these people's lives.

Even then, the sleep of such people is very troubled and the influence of the worries that are constantly emanating from their head may even overpower the dream time. It is a sorry situation that can only be turned around by addressing the possessions that have taken on an overpowering influence in one's life.

To simplify and assess your possessions is certainly essential at times to even open a door to the inner depths. This takes great courage and strength. When their whole life is dedicated to serving these possessions it seems to the person that if they should change their life style and free themselves up from being slaves to their acquired things, they will have nothing left. It is as if they will fall off the Earth because of a lack of meaning, for their only meaning in existence comes from acquiring a life style that is expected of someone in this society who is a success.

To give these possessions up seems to be such an ordeal for it anticipates a life of remorse and lack of self-worth. This is exactly the case when self-worth is equated with possessions rather than a spiritual sense of who they are.

It is devastating to look at a world where all have bought into a life style focused on material goods and any alternative way of living seems obscured by the blinders they have acquired. It is only in the salvation offered within themselves that the necessary courage and fortitude can be found.

Here, within themselves, is the spiritual basis needed to

help them move to the new path, yet in order to hear the voice within, it is necessary to change the external focus and this focus is demanded by the possessions one has. It is a dilemma for any person.

If they decide to simplify their life to hear the inner way, they must find the motivation. And to find the motivation, they must simplify their life. It is only by listening to the faint voice coming from a long forgotten heart that a person can have the fortitude to resist the temptation of a commercial world and take the necessary plunge into their own inner realm.

We can only help this effort by assuring these people that the riches of themselves await them within. If they take the plunge, the blinders will be removed from their eyes after they have found the courage to go inside of themselves and listen to their own heart.

It is here that they will find love and comfort. Material riches can only inhibit their growth and evolution. It breeds a life of stagnation which is opposed to the flow within the cosmic realm.

We encourage all to look at their lives and see how their possessions are inhibiting them from seeing the world for what it is. A place of spiritual growth can be attained by listening to the inner messages that are freely given each day to help you acquire your true possessions - the gold that lies within. For it is in this place that your self-worth lies.

Material possessions decay and require constant attention to keep them the same. The attention you can give to your inner worth will result in growth far beyond your expectations. It is here in the place of your heart that you can hear the drum beat of your destiny.

Forego the shallowness of a world that seeks only the

maintenance of the machines of wealth. There is no greater reward than to find out who you are in the great scheme of evolution.

46

It was not until morning, after a night of little sleep to bring consolation to his turmoil, that Allan was willing to step back from his previous behavior and look at where he was. Again he approached the mirror in his cabin to stare at his face, and in spite of the turmoil in his thoughts, he saw a relaxed face fully content to be who he was. And as he gazed on this face, he recognized how his body was content to be in the flow of the great cosmic movement and he relaxed into it.

As he relaxed, it was as if the avenues of his tormented mind looked around and realized the past was just that and it was time to recognize where he was. As his mind quieted and began to feel the flow of living energy that pulsed through him in this center place, he could see the love that flowed always to his center and he realized consciously for the first time that there was no separation between him and the universe.

All his torment was stuck in its own dilemma without realizing that it all had one basis. And now he saw it. His self-esteem issue was a constant search for the love he knew he needed. Rather than searching inward as he now was doing, he looked outward hoping it was on the faces of the people he met and in the roles he sought for himself in the ways of society.

He saw this outward search for self-worth as fruitless now, for he felt the love of the universe surround him. Allan recognized himself as essential to the whole, and in the whole of the universe there was no other place that was like him, for all was unique and special. And in the broader view of where he was and who he was, he received the recognition he needed from the center of the universe as it wrapped its flow around him in an embrace of recognition. It was as if the eye of the center looked at him from a mirror within him and told him what he needed to know - that all was interrelated in a sea of love that enfolded all the living essences in an awakening of consciousness.

He calmed down seeing that the flow would sustain him in this time, and the feeling was not to dissipate for consciousness was there to sustain it. And in the light of consciousness, he reinforced his own receptivity, saw it in a different light and marveled at the world beyond the world of the physical. For in the swirls that danced in the far reaches of space far above him, there was an intimate connection with his own center that eliminated all the space in between, bringing him close to the place of the true center - a center that pulsed not out there, far from here, but next to him. It was so near to him that where one left off and the other began was not to be seen or felt and as he plunged into the All nothing could now substitute for this belonging he felt.

Allan moved to the couch and sat down feeling this sense move with him, feeling the whole in him beating its beat, pulsing in loving tune with his new time sense. He sat there for many hours without movement wondering why he had searched so long in the wrong places. And as he sat there, he knew he had reached the core of his being, and there was nothing that needed further clarification, for he could now do the work in the way that was appropriate. Time and space had collapsed into a brilliant spark within him that set off his own glow which permeated through his inner center, always sustained by the movement that pulsed from the Center. Time meandered in his quietness and all waited as he remained in the place of sourcing, for at times like this all activity ceased in expectation of the new birth.

47

We will continue on with the lessons and talk about relationships.

There are many relationships in the physical universe that your scientists have categorized as various physical forces that keep bodies in space moving. This is, of course, only a surface view of what goes on. To see the larger picture one must delve into the forces that are within - within the universe, within the galaxy, and within the planet and her people.

We do not speak of "within" as the place in the space between solid bodies. We speak of within the essence of that which lives.

All of the universe is alive and this All has an essence that moves and vibrates with a rhythm that affects all the members. The word for this movement and pulsing is called, for lack of a better word, evolution. It is that movement which seeks to grow in a spirit or spiritual sense, constantly growing to become more of who we all are.

These words are confining at times but we will try to speak as clearly as we can.

Movement is constantly going on. It is never a movement that is independent. Everything in the All is interrelated. The

All moves in harmony with the movement of the essence of each living thing, yet the All moves to express what is happening within the essence of the All.

The essence is not independent from the essence of each living being. They are all interrelated. Nothing is separate unless a living entity chooses not to be included. This is devastating to the All and devastating to the one who chooses the separation, yet it is allowed for movement is a choice.

When a living entity realizes how it can participate in this dance of evolution, the choice is to participate. When a living entity is not aware of this living movement of the All, then the choice to be independent seems logical.

Humans have made this choice for independence and look upon the movement in the cosmos objectively - that is, they see it as something that goes on outside of them. They view the movement as something that should be studied objectively, having no relationship to their personhood. This leads to false conclusions about the nature of the universe and the nature of man.

All forms of life have an expression of their essence. The planet Earth expresses herself to the universe and is receiving the essence of the All. They move in harmony with each other. And the planet Earth expresses herself to her children and most receive these messages and participate in their own essences in harmony with her rhythm. They are inextricably joined, yet express their unique nature.

However, humans are not participating with her. They view this home as an inert body that is governed by physical forces. This is not true but they persist because they wish to control the relationships in which they participate.

All human relationships are governed by laws that humans have created without addressing whether these relationships are appropriate to the cosmic evolutionary movement. They see themselves as independent from the essence of things including their own essence. This can only lead to disaster.

Many see the Moon as also a physical body that exerts her influence over the currents of the Earth's water because of her external movement around the planet. This is only a surface conclusion about what is happening. If you would only delve into your feelings when you think this, you might see what is going on.

You feel the Moon's movement if you are awake. The Moon calls forth life cycles, not in any objective way, but as a response to the Earth and her life - all her life including humans. There is a very intimate relationship between the Earth and the Moon.

And there is an equally intimate relationship between Earth, Moon and Sun with the living growing force. Yes, you say, you understand this because in a way you feel it. And we say then, why do you objectify it?

Why do you see with one eye the feelings you feel between yourself, the Earth, the Moon and the Sun, yet only see them as physical bodies with your other eye? Why does this make sense to you? Because you do not delve into your inner depths to feel your living essence pulsing in the dance that you see all around you.

You see the relationships among things to help you realize what is going on in your universe, yet you see them as something out there, an objective universe that has no feeling to you. You only feel it all when you are not so objective, having dipped shallowly into your own self. Imagine what could happen if you went all the way in.

Such are the ways with relationships. They are within your own essence to feel. There is nothing that is separate, or objective. Can a scientist disconnect himself or herself from what is being studied? The answer is an emphatic "no." The nature of the universe is not separate bodies governed by physical forces. This is a shallow view that does not see the living essence of the All. We say this to show you how shallow are the relationships you have formed in an outward way and to show you how much more powerful are the relationships that exist within you, within the Earth, within the galaxy and within the All.

48

The morning erupted on the cabin in the woods where Allan slept. He had opened to a place during the previous day where consciously he could feel the pulse of the Center beating in him. He knew that his worth came not from the external world but from the sense of belonging he now felt in his heart - a belonging that was larger than he could express for it was beyond even his comprehension.

Yet, it was not without acceptance, for he had a knowledge that accompanied his new sense of worthiness that eliminated all doubt about the scheme of things that permeated his being. It was as if he knew his rightful place in the movement of evolution and in this place he was further motivated to be the messenger he was called to be.

All his doubts and concerns about going back to the world faded, and he saw his own steadfastness in his new role. Yes, it was not easy to bring such a sense to a world that could not see beyond the physical proof that was elicited in experiments and data, but it didn't seem to matter. What he felt was truth far beyond what science could deliver and he knew the spiritual sense he now contained was all the strength he needed to face the public again.

As he awoke, this new found self-worth was still there,

beating in his heart and he thanked the spirits who guided him to this place. Yes, there would be further changes for he could feel his spiritual body affecting changes in his manifestation that would go beyond what he now felt, but that was okay. There was no rush. He only wanted to move in the flow that emanated from the Center, and he knew all else would be shown to him.

He moved to the front of his cabin and watched the sun rise out of the East and in this rising he saw his own awareness grow again, realizing that all could reach this awareness if they but listened to the place within. He hoped that his own story would help the process and he looked forward to sharing his experience.

And as the light of the morning took over the scene, he walked softly back to his couch, sat down, and paused before any further preparation for his move. He continued to feel the strong pattern around him, the flow of living energy of which he was a part and he marveled at the way of this place that had become conscious within him. In this light he accepted who he was without fanfare or turmoil but with a quiet knowing that all would flow in its own good time. And the creatures of the woods rejoiced with him in a hope for the future that looked bright and fruitful.

49

We wish to talk about the way humans respond to the messages that are available to them.

It is quite a shock for a human to realize that there is another part of themselves that knows more than they do. Humans pride themselves on their great intellectual capability and delude themselves into thinking that they or other human advisors know what is going on.

To recognize another reality that has more wisdom than they do is a challenge to their own self worth, for their worth is based on their intellect rather than their spirituality. So, many times there is a great reluctance to open to the other world where information about themselves is readily available if they could but listen and understand.

Recognition of this other dimension of themselves puts many things into question. It erodes the underpinnings of their externally focused way of life. It calls into question the way they have established to teach young people about who they are. It calls into question their total focus on external growth while their own inner growth is stunted.

This realization is quite a challenge and the immediate reaction is to reject all avenues that lead to the inner way. They try this until the cumulation of garbage heaped onto the inner garbage dump reaches such proportions that they feel distraught and lost to such an extent that they will do anything if it can just take care of that desolate feeling.

Yet, as they search for meaningful ways to heal themselves, they again gravitate to anything that relies on external methods. It is not that these spiritual methods are faulty in themselves. They could help humans over the hump, so to speak, if they were combined with inner work. Yet, it is this reluctance to going inside that keeps the human searching for remedy after remedy, all to no avail.

They are willing to work all hours of the day to accumulate external wealth, or to accomplish some external project, but they are reluctant to repackage their day to allow some time to work in an inner way with their dreams. It requires a major trauma to encourage any changes in life style such as a heart attack or cancer or any of the other symptoms that indicate major inner problems are happening.

Yet, the life changes only focus on the external way - ways to eat better, to exercise better, to try and release the stress they carry. And when they are encouraged to look at inner ways, they see them as relaxing methods rather than receiving the messages that can help them to heal.

We say all this to try and reduce the delusions in humans. They see their way as the only way and wonder why this or that method doesn't help or why it fails after a time. We tell you that all you need do is open up to the inner way - to listen to the messages that come from your own depths and begin to understand them.

Yes, it is a different way and it is difficult at first. Change for this generation of humans is difficult for they have built their whole way of living on the external and believe that anything can be fixed in an external way given enough effort and concentration.

This is not so, and will not be so. The mess that humans have made in their external world is only a symptom of the mess that goes on inside them. It is here that effort must be placed. Otherwise, the lifestyle of humans will lead to the eventual disappearance of their species. For it is in the nature of humans to tend the inner garden and if they refuse to follow their nature, they will atrophy.

Great effort is required to reverse this trend. Great effort in an inner way. If you ignore the messages that come from within, you ignore your own salvation. We can't stress this enough.

The rewards for doing inner work are great, but the devastations in not listening to the inner messages are equally great. Realize that you are lost and cannot see the solution and realize that you are spiritual beings who have the great connection to the All.

It is not your reluctance to see that keeps you from changing, it is your inability to see the truth about who you are. If you would only understand the great love that is within you, you would not be reluctant to listen but welcome the opportunity. But, as it is, you fear what you may learn and choose to go on your same path.

We have only this to say to you. The cosmic realm reaches out to you to participate in your own spiritual movement. There is a great need in all of you to take on your mantle and live out of your own spiritual being. This being resides within you and you only need to open the door to the inner realm and the wherewithal to accomplish the inner work will be given. It is your destiny, one that encompasses the All. We all await your response.

50

Allan spent the day moving slowly for he was in a process of consolidating all that he had learned in this place unto himself. He mused about his journey here to the cabin, drawn here by a dream. And now he could see the changes in his life taking shape to change all that he came to know about the world.

It was a time to see with different eyes far beyond a life that was motivated by things within society. This was no longer to be and he marveled at how the turns of one's life could bring such dramatic changes.

He loved these woods he lived in and had little to do with the community here, going into town only for his necessities. Now that was all to change for he was called to move to a mobile lifestyle. He knew this meant being on the road, being able to meet more people, but he was reluctant.

There was a peace here in nature that comforted him and he was unsure of his own ability to hold his center as he moved into the activity of his culture. Could he insulate himself enough that he would not be sucked back into the rapid pace of activity that everyone seemed to love?

He did not know for his new found strength seemed tentative as he thought about the ego energy he would have to face - a potent power that could sweep him up into the daily concerns of a physically oriented world that knew little about the unseen realm.

51

We will begin today with a message to all those who struggle in their day-to-day lives trying to make ends meet. They wonder how they can accomplish anything spiritual when the world is falling down around them and they are busy trying to prop up the supports that keep them alive.

We say to these people to get out of the system that entraps them. It takes courage to look at other possibilities when you are so entrapped that you cannot see beyond your immediate concerns. You worry about providing for your children, having a reasonably nice place to live, not looking so derelict that you are noticed for one who is not making it.

You want to have dignity yet the cost is high. Realize that you equate your dignity with what others can see about you. If you have shabby things, then you allow others to make you think you are like the things around you.

This is not so. You have a great strength within you that can support the changes you need to make. Do not get stuck to a physical place. Do not fear moving from your present location. Get rid of the shackles that keep you planted in this place which brings nothing but misery.

Be willing to look at alternatives. Be adventurous. There are many places on this planet that can support you. Do not think that because you only know one way of living that there are not many other ways that can support you in life.

Believe that you are loved by the Creator and, if you allow it, you will be led by the voice that rises within you.

Do not worry about your children. They are more resilient than you are and they love and honor your inner wisdom because they can see it more than you can.

Listen to your inner voices in the dark of night and allow the voice within you to enliven you to all possibilities. Do not lose the incentive that you get in the dark of night when the sun comes up. It is real and the voice will lead you.

Do not fear that desolation that you see in front of you for you will be guided if you simply believe in what resides within you. Forego your concern for the material trappings that enslave you to a way of life that does not provide any long standing support.

Life is not about the material but about the spiritual. If you live from the inside to the outside you will be surprised at how much of what goes on in your daily life can change and bring you the strength you need for movement.

You live close to the Earth, not in lofty ways of the head that separate you from the movement of Mother Earth. You are close to your roots if you only allow the voice inside to be heard.

You do not need to seek dignity for you already have it within you. It is your own inheritance that resides within and awaits your vision of what is there.

The time has come for all to see what has not been seen and to hear the voice of the Creator within you. Be attentive and do not limit what comes up from within. Nothing is impossible if you listen to the depths for you are close to them.

Let your despair uncover your goodness. Let your emotions express themselves and lead you to your heart. This is the place where you can find your own passion for living. It is a place filled with abundance if you'll just have the courage to let go of what you have been taught about how to live, and reawaken the spark that can lead you to new ways that others have not thought of yet.

Remember, you are close to the heart beat of Mother Earth and can hear her call for you. Have courage and don't fall into the trap of enslavement. Be free of the do's and don'ts and believe what's in your heart for you can hear what is needed at this time.

You live close to the Earth even though you cannot see it. Be with your feelings and let them guide you into yourself. Do not reject the deep feelings for they bring you close to where you need to be. Fall into your depths and allow the Earth to guide you. It is your destiny to be the fruit of this land.

52

Allan went to sleep that night troubled, wanting to delay his move. He was fearful of what was ahead, not so much about doing his mission, but fearful about holding his own Self above the turmoil that he was about to enter. And so he slept a fitful sleep worried about what to expect.

His dreams played out his turmoil, with the chaos of himself portraying him as a frightened animal fearful of the cage. And yes, it was true, he thought as he awoke. He did fear the cage for he felt society as confining, but his old concern about being considered crazy reared its ugly head.

He saw the cage like an animal sees a zoo - a place where you are put on display because you are different. How silly that was, for the public never saw the wild animal as he was, in his natural habitat, no matter how well they tried to simulate it. And the animal paced back and forth agitated for he knew no where to go to escape the staring eyes.

As he made breakfast, he could feel the eyes on him and he knew the fear was back. It was a fear of being looked at as a wild animal that could only be a curiosity in a world of machines and concrete. Where had his confidence gone? How would he make the move back to society if his inner strength could dissipate in an instant?

He rushed through his breakfast and sought the solitude of the woods. He tramped through the brush in angry steps, steps that no longer walked softly but cut through the woods in a destructive way. The life in the woods stopped to watch, wondering about the turmoil they sensed in each step. They could see the turmoil around his center and wondered why he left this trail of destruction when he only had to stop and see where he was. Curious, they thought. Such a creature this human was to contain such energy.

But Allan did not hear them for he was caught up in his own apprehension - a made-up world that confronted him, built in the confines of the head that saw only the disaster of what would befall him. He walked aimlessly, listening to the voices of this or that scenario that would confront him after he left this place and in his thoughts he could not find that place of peace he had experienced the day before.

He finally returned to his cabin ready to ask for help. Returning to the couch, he asked the Guardians to tell him what he was doing to himself and, of course, they did in no uncertain terms. And he began to see what his anticipation was doing to him.

In his own mental fantasies he had created a world that charged with the outrage of the enemy about to be attacked, as if he would be noticed instantly as he came into a city. He gave too much credit to a society that was so involved in their own race. One human who was different would not elicit an anger that was life threatening.

No. This was not to be, as the Guardians calmed him with their voice. There would be no recognition at first, for it was not a world that was easily distracted from its own goals. The anticipation that he felt within him about returning to this world was played out in his mind as an instant of time. That would not be, for the pattern always offered transition time and so it would be for Allan. There would be time to grow accustomed to the pace of the world, time to see his own pace and time to approach the few who would listen. It was not to happen in a blaze of fire that would transform the world.

No. It would be gradual enough to allow him time to deal with his new found self-worth, to feel his strength as he walked, and to be willing to talk of his experience. He calmed down then, realizing his stupidity as he built up the fantasy of his own apprehension into a time frame that would be played out in one moment to crush all that he had worked so hard to attain.

No. He saw that now. And he calmed into the words of the Guardians seeing the mental images he had created fade into the distant scene. His feelings of being caged were dealt with in the reality of his own Self as the turmoil dissipated and his calming point of the center settled over him.

Time. No longer an enemy. For it was time that would chart out a path of transition, of growing activity appropriate to his own acceptance of where he was.

Time. Once an enemy that he always had tried to overcome, now was an ally that calmed his fears and assisted his growth. How strange this felt to the one with wings on his feet, enjoying the fast pace of achievement, demeaning others who could not see as fast as he. Now he was welcoming the slower pace of the center, relishing the calm provided by days where he could become accustomed to his new way. He had time to feel the pulse that beat through him, a beat that beat the rhythm of a universe that moved in ever changing patterns to fulfill the movement.

All calmed in him as he embraced time as a friend. And he relaxed, knowing he would be brought along at a pace that was appropriate. The only hurry was in his head as he ran from a threat that wasn't there.

Yes, he knew the threat may materialize, but only at a time when he was ready for it. For now, he would be given time to grow accustomed to the center place even as he prepared to travel. And the pressure valve opened releasing his fear, for he had dealt with it in the only way of release - in the growing of consciousness, in the understanding of what he feared, and in seeing it in the strong light of meaning.

The cage he feared was of his own making as he anticipated what would happen. He realized the wisdom of a saying the Guardians had given him a long time ago - don't anticipate, participate. And he saw the flow of time spread before him, allowing him to participate in a journey that would be carried out at its own pace - a pace that would be appropriate to the movement of time that had opened up in him in a new way.

He saw now that he had finally passed through his nemesis and had reached the other side where enemy turned into friend, and dark turned into light. Allan marveled at the labyrinth that had led him here, to a place where time and space were opportunities rather than threats.

53

We wish to discuss the way people seem to hold onto their own reality that this world creates and deny the existence of miraculous changes within them.

There is much to be said for keeping one's head above the water and not clouding judgement with delusional fantasies that inhibit clear thinking. Yet, the human species goes too far with this concern and relegates the inner world to this fantasy world because of the way this world communicates - in an unseen way with dreams, intuitions and meditative states.

There is also the fear of the bigger picture that is presented. It is difficult for people to accept that there is a higher purpose to life. So, all this is also put into the fanciful category from which people wish to stay away.

Let us look at this with a clear eye. Delusional thinking is thinking that fantasizes about what is happening to a person in their everyday life. They find themselves dreaming during the day about becoming a fashion model, or becoming a movie star, or becoming some famous inventor. All fantasies about improving their lot in life.

This is not bad or detrimental to the situation of the person until it starts causing unusual behavior like acting out the fantasy without having attained that goal of being the person they perceive they want to be. It is forced behavior that eventually may cloud the reality of the person and they begin to believe in this forced behavior or face they present to the public. Yes, this can be very detrimental for it will in all likelihood move the person further away from their own inner sense of who they are.

Now, the inner realm is unlike these fantasies in that this realm is the closest to who they really are. So, when a person begins to receive messages from this realm, it is vital information. It is here at this point that the person needs to distinguish between the daydreams of their head and the messages that come from the inner realm.

At first, it is best to rely on dreams because the ego has the least ability to stop these communications and interfere with them for the ego is asleep when they occur. Thus, the ego's fantasies do not enter into the picture.

Another benefit is that the dreams do not feed the ego's mind games but try to bring reality into the picture. This can be seen by the uncomfortable way the ego feels in seeing the truth of the messages that are presented.

Intuition is difficult at first to distinguish from the ego's influence in painting its own picture of reality but if the insight comes without warning, it is generally true. In other words, intuition is true if it comes upon the person sort of "out of the blue," without any thinking about the subject presented beforehand.

Meditation can also be judged as truthful if the person has allowed the meditation free reign to meander in the inner realm. If there is any sense of control within the person during the meditation, then the images or messages received may not be unencumbered with the ego's desire to achieve certain results.

So, at first dreams may provide the best means to listen accurately to what the messages are, provided the ego does not interfere with the recording of the dream or doesn't try to interpret the dream in a way that ignores the truth that is presented.

We tell you all this to show that the culprit in seeing the world differently from what it is, is usually the ego and its own desire to maintain control. This maintenance of control of what the ego wants to achieve can become an entry way to creating the fantasies into which people fall. It can come upon a person when it is least expected because the ego is quite adept at building false pictures of reality.

So, we tell you it is not the inner way that will create these false pictures but your own ego in its constant attempts at controlling life and presenting pictures of what could be to entice you to act in certain ways. The inner way seeks the truth. While its language is highly symbolic and may seem unworldly and fanciful with its images of castles, snakes and actions that are impossible in your current reality, it speaks the truth if you but learn the language.

So, do not stay away because you don't understand the messages and think it is fantasy. This only leads to fantasy for the ego then has no balancing force to counteract its own meandering of the head. While you think you are keeping close to reality by rejecting the dreams and other inner messages, you actually move further away from reality.

Talk to people who know dreams and see what they see in the dreams and decide for yourself if this is reality or some fantasy that is going on within them. Listen to them closely and see what they harvest from the inner way. It is true there are not many today following this way, so you may have to look long and far to find a practitioner of this method, but your search will be rewarded with a new sense of reality.

We tell you that your concern about keeping your head above water is misplaced. Look to your own ego - your conscious part of yourself - for the one who can create the most outrageous fantasies and can lead you to a place where you have deluded yourself.

Allow the inner messages to come forward and see yourself as you are. There are more realities in the world of which you are currently unaware, and some are more real than you care to believe at this time.

Be open to your inner realm and you will see with clearer eyes and clearer thoughts. Then you can decide what is fantasy, and what is reality.

54

Allan had made a tremendous leap in his own journey for when a person has dealt with his or her ruler, in Allan's case time, and come to the other side seeing the positive aspects of this, they were then allowed to move beyond the physical constraints of their body. So it was with Allan.

He had not come to know where this would lead, but he had begun to have sensations that were not readily explained - sensations in his body as if various parts were suddenly the recipients of some strong energy that temporarily caused some discomfort but then quickly subsided.

And his inner sense also made some shifts. No longer were dreams confined to the deep sleep, but they also began to be manifest in the middle place between waking and sleeping. He began to see images at these times and he wondered how this could be.

These changes had been taking place throughout his time in the cabin, but his own awakening of consciousness moved him to a place where understanding of these higher realms could be attained. He had also moved his consciousness of what was happening physically to him to a place where he was more aware of the subtle movements that were even now transforming him.

He also began to remember more of his spiritual nature. This was quite disconcerting for a human to remember such things because this realm had always been relegated to a place of fantasy. It was as if the mythology that had survived the manipulation of the human species began to come alive in his new understanding. And this understanding did not relate very well to the physical world, that world that seemingly confined the living world of humans. There was more than meets the eye, so to speak, and Allan was beginning to catch a glimpse of a world beyond his comprehension.

55

We will continue along the process of the inner way. When a person begins to approach the sanctuary of the center place, many new feelings begin to surface. There is a movement from the mundane into the more cosmic or universal realms where the symbols take on a more mythic character.

At this point, it is important for the person to reflect on the view they have of themselves. There is a change going on that transitions the vision from the outer world, where one manifests the inner person, to an outer world that informs the inner, just as the inner informs the outer.

It is a time when the barriers between worlds get hazy and what were the symbols of the inner world begin to be seen in the outer world. It is at this time that the person must be in a heightened state of awareness for all that is offered.

This is to say, that as the person approaches the center, the separation of the mundane outer world and the symbolic inner world fades and they begin to become one and the same, both potent in the messages that are given. At this stage, the person must apply their new capabilities to understand the inner messages as portrayed in the outer reality.

The time when this happens is up to the individual, but it

can be realized when the symbols in the outer world start to become filled with potent messages. For example, an opportunity is offered by a passing friend; a message is received that when followed creates new opportunities; an opening is established in some new line of work that has the power to change the person's life.

It is here that the person needs to see the symbolic significance of all that is happening in the outside world. This does not diminish the dream time but adds another dimension to the journey of the traveler. The person can now begin to see that the inner is reflected on the outer, and the outer world becomes a drama that unfolds the inner. All separation is blurred and it is at this time that the human begins to feel the broader vision creeping in.

Before this, the inner world was just that - a different, symbolic place that somehow gave messages to help with their life, but was safe behind their closed eye lids and could not intrude into the waking life. Now, once the vision clears, this is no longer the case and the world broadens to include more inner and outer together, joining in a harmony that is at first difficult to comprehend and is frightening to the initiate.

The world of symbols becomes real and powerful at this time, and the real world takes on a glimmer of the fantastic, as if the entire universe were concerned about each step of the individual. This seems so outrageous to the individual that there are many doubts about what they are doing. They begin to feel as if they are moving into a world that is definitely not real and they should take a few steps back and reevaluate this whole inner way.

We tell you now that there are many doorways you must go through, but this is the most difficult for it challenges the whole way you see the world. A sign on the side of the road may be symbolically significant to your journey. Something wrong with your car may be telling you about your vehicle of life. A purchase that you make may create all sorts of problems causing you to see whether you need this item and what it all means in terms of your journey.

The process at this time is a difficult passageway for it brings the human's view of reality into question. To transform into this new reality requires a strong person with a belief in the sanctity of their own journey, for the fear that this change will transform the way a person sees is well founded. It can only be taken on by those who are willing to see the larger picture of life.

Once this transition is accomplished, the human can no longer be seen as separate from the mythic realm, but must recognize that life is a potent journey of awakening that includes the All. For the concern that each step in the human's journey is of vital concern to the universe is real. This is how it is.

Your life is of vital concern to the universe and all of humanity. At this stage, you will begin to believe this and because it is so unbelievable to you, you may hesitate to enter the center place. We can only tell you that the hesitation is normal and gives you time to realize the great worth you have.

Hesitate yes, and contemplate the full meaning of all the feelings you now have as you connect to the All that is. Step forward and you will enter a world view that you never knew existed and once there you will see that it is the place you have longed for your whole life.

Take time, for this transition is a leap of faith at first. We encourage you to take the plunge if you know you are ready. Never force it. If you feel you are not ready, listen to your dreams and believe what they tell you. At this juncture, only your inner group can tell you what you should do.

56

Allan continued on with his chores to get the cabin ready for sale, for he knew he had completed his work in the woods. Yet, his expectation of how he would feel at this time was not related to the reality of it, for if anything he felt ill-prepared to become an emissary for the clarity of his vision had become opaque rather than crystal clear.

This bothered him and he began to have some conversations with the Guardians about the state he was in. It could only be described in his vernacular as feeling "out of it." And, of course, this was an accurate description for he had moved to a place that had moved himself and his vision into a world that encompassed the day-to-day world but also partook of the spirit world. This could not be felt as the usual normal state.

He began to have dreams of the number three in various forms, always depicted as three separate things coming together in a way that the third was an outgrowth of the two. He had worked with this symbolic content before, realizing that the three was the synthesis of the two opposites - a new creation that emerged from the tension of opposites that had been born in him.

He had been searching for that part of him that was the

new creation, yet it had eluded him. And so, the private sessions with the Guardians began to help him see what was happening. And his dreams began to make sense as the patterns of symbols moved him forward to his own full awakening.

As he was led gently to his own conclusion, he began to understand that his role, his mission that he had come to accept was the synthesis and he was shocked. Not that it was not understood, for it was, to the depths of his soul. No. What shocked him was that he could not see it, for it was directly in front of him.

This was not surprising since the most obvious became the last thing one looked at, for it was too close. It was like a person landing in some foreign country looking around to see where it was without taking the time to see where his feet were. It was so close to the person that he looked in other places to find it. Such was the extent of Allan's looking. He looked for the result of synthesis in the future where it was to happen, rather than in the present where it had already occurred.

And so Allan was shocked. He needed time to come to grips with the realization that as a messenger he had become a new creation - something beyond the opposites that worked within him, beyond the combination of unconscious and conscious, of dark and light, of receptive and active, of all the tensions that moved within him. These opposites now had propelled him to a new place that was the place of physical and spiritual but reversed in a sense, for the spirit world was now the primary realm that encompassed the physical, rather than the physical partaking of the spiritual.

Yes, it was a subtle change but a change that began a transformation of vision and physical sensations that brought

Allan to a place where nothing seemed the same, yet the day went by as usual. It was a place where his understanding was growing into a form that encompassed a feeling of subtle differences adding a sense of "other world" to his normal experiences - a feeling that he was partaking of a world beyond his normal world. It was a world that encompassed much more than simply understanding the broader vision, but a place where he was moving into it.

It was disconcerting to him. His cabin was there, and his chores that needed to be done, and the woods were there outside as well as the community that supplied his needs. But there was another feeling that crept over him in his every movement - a feeling that he moved through space and time with a different sense. Nothing it seemed could describe it, other than the sense that he was part of the physical but also partook of a subtler realm that lingered in the back of himself, in his periphery, that almost overpowered his own sense of reality. It was like a different world was attempting to enter him and add to his repertoire of senses.

And so he carried on as he had planned trying to deal with this growing "other sense" that crept closer. As it approached, he tried unsuccessfully to hold onto his body as his vehicle, for his position out of which he worked was changing quickly.

At times, in the dead of night, he panicked in his dreams and tried to fight with these enemies that crept toward him. But he could not distinguish who the enemies were for those that approached him did so in such a subtle way that the threat was not clear. He began to see that the only way was to allow the changes and this eased the transition slightly.

57

There are many ways to be in communion with the cosmic realm. We have talked about dreams as a true guide. We know this is difficult at first because of the symbolic language that dreams use, yet it is really the only way that your inner depths can talk to you.

At first, most cannot communicate directly with the inner realm. Some never have this talent. So, direct communication is not very reliable for those who haven't spent many years working with their inner depths or for those who have no talent for this.

Dreams provide a universal way for anyone to be in communion. Yet, the language must be learned in more of an intuitive way rather than taking classes. This way of learning requires experience and it is best for people to just plunge right in realizing that the things and actions seen in dreams have to be translated. For example, dreaming of a bear requires knowledge about the bear and its affinity for the dream time in its long hibernation.

The meaning of symbols has been acquired over many millennia and there are sources that can be used to understand these symbols. The house, for example, has been known from almost the beginnings as symbolic of your inner realm: the basement as the unconscious that you don't know about, the kitchen as a place of transformation, and so forth.

So, it's best to talk to people and find the sources they use. Remember this - a source must deal with the ancient symbols, for the dream time is from the timeless realm and there is a consistency in the meaning of some of the more powerful symbols like snake, fire, water and other constant things that have been around for years.

Of course, some modern symbols are picked up by the dream time to help you see more of what is going on in yourself. Symbols such as subways, railroads and airplanes occur in dreams and clues to their meaning can be gleaned from what you know about them. Subways go into the underground or the depths. Railroads travel on tracks and are powerful ego things. Airplanes travel in the air so knowledge of the element of air is important.

Today there are many resources, so ask people who know dreams what they use. It is at first confusing, but if you begin then help will be provided.

Remember that dreams talk about you. Many times you will see images of people you know in the outside world. Realize that the views you have of these people are parts of you. For example, if you see a friend who seems to have a talent for organizing things and you admire that, realize that you have this capacity within you, even though it seems far removed from your view of yourself.

All these people you see in your dreams are part of yourself and you are not aware of them. In fact, you might look at this as getting in touch with the people that are contained within you. You, the conscious part of yourself, are getting to know those aspects of yourself that are presently unknown.

There are male figures who are active parts of you and there are female figures who are more receptive parts of you. The male figures help you to manifest actions in the world; the female figures help you to receive understanding and guidance from your depths.

At times these figures may be opposed to your journey and they then would be parts of you that are not helpful. These are parts of you that hinder your development and you will have to face them and confront their attitudes.

All this goes on to help you know who you are. So, it is not easy work. You have to work with the symbols and see how they relate in the action of the dreams that you have.

Remember that the dreams give you true messages that have not been filtered by the conscious part of yourself, so be very honest when you remember the dreams even if the images and actions you see are troublesome. Don't edit what is given. Work with them as they are and ask for help. You will be rewarded with great insight if you persist.

It takes time, for that is the purpose of time - to give you a chance to peel away the layers that surround your center place, a layer at a time. You could not take the entire story in one sitting. This could overwhelm you and might have detrimental effects.

The dream time gives you what you can handle at the time. It all relates to what is happening in your outer world for this world is tied to time. So, the timeless realm and the realm of time work together to bring you further along the path to knowing yourself.

And as you travel along this path, you move closer to the communion you seek with the Creator. Can there be a better goal that you seek? To live in harmony with the universe and find your place in the great scheme of evolution?

You will then begin to know who you are and a wonderful world will open up for you. Adventures beyond your dreams will unfold, and the world will be your playground.

58

Allan had some difficult days as he had begun the process of becoming a spirit being rooted in his own physical nature. He had to be consoled by the Guardians because of his conflict between two realms - the lofty spiritual realm that seemed so out of touch with his own wood-loving nature that was a part of who he was. The conflict of becoming a messenger within a body that acted as natural as a camper on Sunday at home in the woods seemed to be one that was not to be resolved.

Nor was it necessary. This was what Allan had to understand, for if he got caught up in the spirit part of himself, he might have a tendency to lose touch with his physical, down-to-earth nature and this would be a disaster. He would then rise to a faulty sense of who he was, exalting himself over his body, losing touch with his true nature.

On the other hand, if he abandoned his spirit-filled side, he would wallow in pity, for he would have lost the part of himself that excited and moved him in his quest of understanding the mystery of life and the journey he was on.

It was a difficult time to feel both parts of himself at the same time and to allow the new creation to be born. And so, the Guardians were with him constantly, helping to balance his actions and instruct his heart that it was okay to be who he was in spite of the lofty way he saw his mission and role in the world. It was only lofty in the sense that it was different, different from other roles he had seen in the world. And it was lofty in the sense that others would think it was lofty because they were used to thinking of a higher realm in this way.

Allan felt the tension of his own presence as one who lived in the woods in contrast to the loftiness this world would think appropriate for a spiritual messenger. In the world, the lofty ones acted lofty, and the woodsy ones acted close to savage. The world in which he lived could not reconcile the two together, but so it was.

Allan could hold the opposites within him and it was this strength that carried him forward into a new being - one who would return to the world as an enigma to what it valued. This was necessary for it was the truth of who he was.

59

We will move on to the next phase of the inner way.

Once the person has willingly entered the pathway to the center place, they will have to encounter the cosmic realm, if they have not done so already. This is a much different world from what they are used to.

Here, living energy is alive and willing to advise the human. This is frightening to the person at first for it brings in a reality that has been feared by the human's culture.

There is also the fear that if the person participates in this world there will be adverse reactions from the society in general. Of closer concern for those who have families is the fear that they will be looked at with disdain by them.

This is a natural concern, for both society and specifically close family members will react to the change in the person. For the society humans live in today is not familiar with the constant communication that went on between humans and the spirit realm in the past. It is not acceptable today, yet it is a natural ability that all humans have.

In this center place is the connection to this realm. It is portrayed symbolically as the connection between Earth and Sky. It is the vertical connection, which pertains to the spiritual, rather than the horizontal connection which pertains to the mundane realm of everyday living.

Here in the center place is the place where the center Self, or as some humans call it, the higher Self communicates with the All. This communication can take on various aspects. It can be direct communication with the center of the universe; it can be communication with various spirits; and it can be communication with the person's guides who have been with them during their Earthly stay.

It is here where the person can perceive their cosmic mission. All this will be a shock to the traveler, for it expands the usual reality into a realm that seems fantastic in one regard, and frightening in another.

It is fantastic because of all the information that is available simply for the asking. It is overwhelming at first to those who are novices for they do not know as yet how to retrieve what they desire. This comes with time and will be revealed to them if they continue to gain experience in this realm.

It seems fearful because the person questions their own sanity. They say, how can this be happening? And society will respond that it is delusional behavior forgetting that they have been told for many millennia that their God resides within them.

It is now time for the human to see what having God within them really means. For too long they have projected this outside of themselves. It is now time to see that each person contains the universe within them and it is here in the center place of their very self that they will experience this.

Their center Self is a facet of the whole, of the All, just as the inner people in their dreams are facets of the center Self. So, is it so surprising that this facet of the whole has a unique mission to perform? At times, humans can be so confused with their own sense of what the world is that they fail to see what is right in front of them.

Yes, you are a creation that is intimately connected to the movement of the All. It just has been buried within you for so very long that it seems foreign or alien to you. We tell you that this connection to the whole of the universe and all its aspects is natural and rather than being delusional, this position pales the view of the world that humans think is real.

This center connection expands the person into their rightful place - a spiritual place that allows the person to see who they are in all their glory. This is the place where consciousness can touch the very essence of the All. It is the place where evolution is realized as the hope of all who participate in the great dance.

Is it not better to be a participant in the whole evolution than to see yourself as an independent ego not knowing where you are going? This is the place for complete discovery and can transform the human into a creation that knows no bounds.

Here the human participates freely in the movement of the All. Here the person takes on a character that is enlivened by the spirit within and becomes a passionate manifestor of all that is there that is unique to themselves.

It is the place where hopes can be realized and destiny can be unfolded. It is the place where love is at its highest purpose for the flow of this love can bring fulfillment in this life.

We cannot take away your fears of this place for it is a place for the courageous. Hear the call of your heart and enter into your sacred place where everything begins to make sense. Do not fear your destiny, for how can the whole be unwilling to see who it is?

60

Allan continued on, holding the great tension that was within by accepting both sides of himself - that which felt at home in the darkest realm, and in the dark recesses of society, and that which felt enlightened beyond measure in his understanding of what moved the universe through its evolutionary spirals of existence. And in all this struggle, he continued to receive the messages he was to deliver to this world that caused him such strife within his own heart.

Late one evening, after the Sun had set and he sat out in front of his cabin, he contemplated all that had gone on and he reached again the place of peace within himself. He was able to accept what he was given in this peaceful time at twilight where all the conflicts in him could be seen as truth, for they were no longer confronted by the full light of the Sun, and the new Moon did not activate his own inner turmoil. And he relaxed into himself allowing what was to be to be consummated.

In that moment, he was given a sign of the truth of what he was, for over the horizon, not obscured by the light of other cosmic bodies, the Dog Star rose and shined its light directly on his heart. He felt the natural rhythm from the place of his sourcing - a place that was revered for many millennia as the seat of the teachers of humanity. His lineage came up within him to moisten his eyes. He saw again the story of his existence, his birth into a world that did not know him. And he saw his subsequent preparation to again walk among them as one who would be seen as unapproachable for his energy was different and it activated those parts of themselves they wished to remain in hiding.

He knew this time was no different from the previous when he taught the Atlanteans about the inner way and saw them walk away to build their cities, more happy in the external world than in the world of their own depths. And he contemplated how his return would be seen by a world that was fully built in their own image, a world that throbbed with the power of ego.

Yet, he knew that as the culmination of this society was reached, perhaps now they would see the emptiness of it all and would again listen as he taught Miriam and Luthor and Morgana the lessons of the Guardians. And in this he was hopeful.

He sat there in the light of the Dog Star content with the knowledge that he was ready to deliver what was necessary, willing to accept the outcome for he knew there was more to this than he could know. The Guardians did not attempt this contract again lightly and he was encouraged by their presence for he knew they would not come without some hope that the path of humanity could be changed.

His heart leaped in his chest at the possibility, for it was the mover within him, the one who sought life in a sea of constant turmoil that swirled in the bowels of the creation. And the movement reached out for this life that could bring consciousness to the One eye so that it knew itself and could know what the dance was, for in the struggle of Allan it had seen enough to encourage it to move closer to this species that feared it.

The eye opened to see Allan sitting there in its light and Allan looked back in communion now with the full movement that approached this part of the universe. And in the exchange of this glance, the consummation of what was to come was seated in the heart of the messenger. Allan rose and gave thanks to the All that was there for all to see, moving through the space of time to bring food to a world that starved for nourishment beyond the world of the ego.

Epilogue

We will finish today with a final lesson. This will be a wrap-up of all that we wish to impart to the humans of this time. We wish to inform them about their great potential and the great sanctity that they hold.

It has not been long since humans have evolved onto the scene of cosmic movement. There is much that is hoped for in this species and much has been put up with, so to speak, in order to allow them to reach this juncture where the path is clear for them to develop into a species that participates in the cosmic drama.

The destruction they have wrought all in the name of ego development has been much, yet condoned, for the benefits that can be achieved through consciousness are great. But this freedom of action will not be tolerated any longer. Payment is at hand if the road that has carried them to this point in evolution is not abandoned.

We do not speak of a great movement to fix society. We do not address political issues; we wish them to address very personal issues - issues that have been long ignored.

The time is ripe to see what lies within. The look inside is painful at first, but consider this the payment for long years of ignorance. Many have suffered for this and the pain can be more easily felt if this is kept in mind.

There has been much destruction, all to reach this time when the hope of human consciousness can be realized. This is the juncture the human species is at. We know no way else to encourage them other than to point the way to the inner Self. This is who they are. This is their destiny. We await their response.

We have given these lessons in hopes that a great awakening can begin. It is only humans who can begin this. They, in the intimate moments with themselves, will have to remember. It is time for this movement.

We do not plead with them for it is their choice. We simply show the reasons in a language they can understand. The results of their movement will be their choice. No one else can make this effort.

Time is at hand. Movements are being made to right the wrong, to move destruction to growth. Nothing else can be done. We give this gift to the human species with great hope that they will listen and see the wisdom of what we speak.

All is in preparation. There is nothing that hinders the movement, only the reluctance of humans to take the plunge can prevent what is now set. So, look in the mirror of your soul, humans, and see the past for what it is and look to a future that holds your salvation.

The choice is now in your hands. It has been clearly stated and there is no excuse that you have not heard, for it is said in your way for your understanding.

It is time to be open to what lies within. It is your choice. May you be prudent in your contemplation.